Advances in Information Security

Volume 65

Series editor

Sushil Jajodia, George Mason University, Fairfax, VA, USA

More information about this series at http://www.springer.com/series/5576

Mauro Conti

Secure Wireless Sensor Networks

Threats and Solutions

Foreword by Luigi Vincenzo Mancini

 Springer

Mauro Conti
University of Padua
Padua
Italy

ISSN 1568-2633
Advances in Information Security
ISBN 978-1-4939-4751-5   ISBN 978-1-4939-3460-7 (eBook)
DOI 10.1007/978-1-4939-3460-7

Printed on acid-free paper

Springer Science+Business Media LLC New York is part of Springer Science+Business Media
(www.springer.com)

Foreword

This book by Dr. Mauro Conti focuses on the most important security challenges for WSNs, conducts a vast literature review on security threats and the currently proposed countermeasures, and proposes several novel mitigation approaches to each of the considered attacks. Moreover, Dr. Conti provides in-depth theoretical, analytical, and experimental discussions on each of the attacks and countermeasures. In particular, a security threat might result in hazardous situation when it comes to WSNs; this is because of their inherent hardware and software limitations in applying traditional security mechanisms, and of their frequent use in countless vital applications.

Over the years, I have dealt with various aspects of networks security, particularly in WSNs, together with my research group at the Dipartimento di Informatica della Sapienza Università di Roma, and with several renowned researchers all over the world. I believe that one of the main goals of adopting WSNs is to provide safety and comfort for humans. Therefore, we need to contemplate how this new technology would guarantee the goals of the designer without threatening the security and privacy of users.

Dr. Conti provides timely information for scholars and researchers desiring to design new WSN systems and applications, to be able to tackle the existing security challenges in these networks. Moreover, this book shines a light for early stage research on aspects related to key establishment, physical attacks, node capture attack, node clone attack, as well as security and privacy issues of specific WSNs services, such as data aggregation. The content of this book is the fruit of Dr. Conti's several years of research effort in security and privacy issues in WSNs, which have also led to numerous papers and patents.

It was my great pleasure to supervise the early research career of Dr. Conti as a promising Ph.D. student, to be involved in all the stages of this work, as well as collaborating with Dr. Conti afterwards. I believe this book by Dr. Conti is a key reference for WSN security challenges, and I hope you will enjoy reading this book.

Luigi Vincenzo Mancini
Sapienza University of Rome, Italy

Preface

Recent technology progress, particularly in the areas of computer networks and hardware miniaturization, allowed the emergence of a set of novel computing and application scenarios referred in different ways, including "Internet of Things", "Mobile Computing", "Pervasive Computing", or "Ubiquitous Computing". Despite the specific meaning of those terminologies and their peculiarities, all those concepts involve the presence of small or tiny devices that communicate (possibly in a wireless way) and collaborate among them to achieve a common goal. In many of these emerging application scenarios, the security of the service and infrastructure, as well as the privacy of the involved parties, is a fundamental feature.

In this book, we focus on a representative technology in this arena: Wireless sensor networks (WSNs), i.e., networks made of tiny resource-constrained devices that have sensing and wireless communication capabilities. In particular, we present a comprehensive approach for building secure WSNs, taking into account different "levels" of security threats: from the basic need of nodes trusting and confidentiality between nodes (via means of establishing secret keys), toward physical attacks such as node capture (physical removal) or node cloning (physically building a new node, cloning the crypto material from an honest one), up to the security of specific applications, where we consider in particular data aggregation, which is a key service in WSNs that can be used to tackle with their energy constraints. Finally, as a representative case, for the data aggregation service we also look at possible privacy aspects, e.g., preserving the privacy of nodes participating in the aggregation—which in practical scenarios might be for example users of smart-metering or other services.

The main contributions of this book can be summarized as follows:

- With respect to the **establishment of pair-wise secret keys** between nodes, we present a new probabilistic solution, the enhanced cooperative channel establishment (ECCE) protocol that overcomes some of the limitations of existing solutions. In fact, ECCE presents higher probability for any pair of nodes to establish a secure channel and a higher resilience rate (i.e., the attacker needs a bigger effort to corrupt the channel). This contribution has been partially published in [46, 47], and is described in Chap. 2.

- With respect to the **node capture attack** (i.e., physical removal from the network), which is the first step for an attacker to perform several other attacks that are crucial for WSNs (e.g., clone attack or the confidentiality violation), we design the first approach to detect the capture of a node leveraging the network mobility—in order for the nodes to trace the presence of the other nodes. The results of our study show that the newly proposed solutions can be practically implemented in sensor networks, and under certain mobility conditions (e.g., a certain average node speed) they perform better than solutions that do not leverage the network mobility. This contribution has been partially published in [45, 49, 53], and is described in Chap. 3.

- With respect to the **node cloning attack**, we first identify the properties that a distributed clone detection protocol should possess, then we design a randomized, efficient, and distributed (RED) protocol for detection of the node replication attack. RED shows better properties and performance when compared to the state of the art. In particular, it is not affected from an important issue that influenced protocols in the literature, i.e., the predictability of the position of the witnesses—hence making the process of detection less effective in practical scenarios. This contribution has been partially published in [50, 52, 54, 55], and is described in Chap. 4.

- With respect to specific WSN services, we focus on **data aggregation security**. The question was to find whether a WSN service can be secure, despite the possible presence of the adversary. Owing to the constrained resources of WSNs, nodes cannot send their own sensed data independently to a collecting point, hence the use of an aggregation protocol is fundamental (and so their security). In this scenario we design the first secure protocol for secure computation of the median aggregate. This contribution has been partially published in [190, 192–194], and is described in Chap. 5.

- With respect to **data aggregation security**, the challenge was to provide privacy to the single node collaborating in the data aggregation process. In many sensor network applications, the data sensed by a single node can be related to a user (or a number of users): Information on patients' health in a hospital, water consumption in a city, etc. Then, in order to protect the people's privacy, the data aggregation protocol that works in this type of context must protect the privacy of each single node. In particular, it should not be possible to relate a given sensed data to a given sensor node. We present the first data aggregation protocol that guarantees the privacy of a node not only against the other nodes but also against the Base Station, which is the entity that eventually collects the aggregated data. This contribution has been partially published in [60, 240], and is described in Chap. 6.

Mauro Conti

Acknowledgments

This book is a revised version of my Ph.D. thesis. I want to take this opportunity to express my gratitude to all the people that made this possible, and have been close to me during the years of my studies. I would like to thank in particular my Ph.D. advisor, Prof. Luigi Vincenzo Mancini, for sparking my research interest in security aspects of computer systems and communications; Prof. Sushil Jajodia, particularly for hosting my visiting period at George Mason University; and all the other people I had the chance to collaborate with during my Ph.D.: Roberto Di Pietro, Andrea Gabrielli, Alessandro Mei, Sanjeev Setia, Angelo Spognardi, Sankardas Roy, and Lei Zhang. Special thanks also to Prof. Cristina Pinotti and Prof. Srdjan Capkun for their valuable comments, which helped improving the quality of this work. Thanks also to Susan Lagerstrom-Fife and Jennifer Malat at Springer, for their guidance during the preparation of this book.

Furthermore, I would like to thank all the great scientists worldwide that collaborated with me at the early stage of my academic career, as well as all the passionate students and collaborators in my research group at the University of Padua: You really make my work so exciting and rewarding!

Last but not least, I would like to thank my family, for the continuous support, and for making all this possible!

Contents

Chapter 1
Introduction

The evolution of computing devices followed different paths. Despite the famous misquotation attributed to Thomas J. Watson Sr., then-president of IBM, ("I think there is a world market for maybe five computers."), during the 70's a new paradigm emerged: The Personal Computer. Computers intended to be used by a single person became so common that the market of personal computers overcame the one of Mainframes. With the introduction of computer networks and the miniaturization of the hardware a totally new paradigm has emerged since the last decade: The so-called Ubiquitous Computing. In particular, this paradigm aims to make "many computers available throughout the physical environment, but making them effectively invisible to the user" [228]. Recent advances in Micro Electro Mechanical Systems (MEMS), in wireless communications and in digital electronic made it possible the production of small, cheap and "smart" devices (that comes also with novel security and privacy issues [5, 6, 12, 17, 58, 151, 164]), such as smart-phones [44, 51, 59, 88, 97, 188, 243], PDAs, Radio Frequency IDentification (RFID) systems [56, 57, 191], Wireless Sensor Networks (WSNs), and many other technologies.

In this book, we focus on the security issues of the representative technology of Wireless Sensor Networks, introduced in the following section.

1.1 Wireless Sensor Networks

In this book, a sensor device is a small device that is able to sense environmental data (sound, light, temperature, etc.) and it is also able to communicate with any other sensor node in its communication range and compute the sensed/received data. A set of these sensor devices deployed in a given area constitutes a network with no pre-established architecture, so called Wireless Sensor Network (WSN). The usefulness of this type of network does not come from the single node capabilities, which are instead very limited, but from the collaboration of a large number of nodes. In a

© Springer Science+Business Media New York 2016
M. Conti, *Secure Wireless Sensor Networks*, Advances in Information Security 65,
DOI 10.1007/978-1-4939-3460-7_1

WSN hundreds or thousands of nodes are usually deployed in a large area where they can sense the environment, compute and communicate the collected data in a very efficient and distributed way. Differently from other traditional wireless devices, sensor nodes do not communicate directly with a Base Station (BS)—a device that does not have the limitations of a sensor node—but mainly with other sensor nodes. So, sensed data are locally computed and forwarded to the BS. The lack of a pre-designed infrastructure implies that each node acts not only as a sensing node but also as an elaborating node and a routing point.

Current and future WSN applications are in different fields [127]: Supporting rescue operation, building surveillance, fire prevention, battlefield monitoring, and so on. Also, as often happens with new technologies, many applications can be designed and thought as far as the technology will be cheaper and widely available. A further description of the possible WSN applications is given in Sect. 1.1.1. In many applications of WSNs, the security of the network is a fundamental issue, as for: Confidentiality, integrity, authenticity, and availability. As an example assume a WSN is deployed for the safety of an area—e.g. for the detection of poisonous gas that could be potentially released during a concert or a big sport event. In this scenario, if the network is not secure we could have a false perception of safety, that can be even worse than the awareness that there is no safety at all.

1.1.1 Applications

Here, we recall some of all the possible application areas of the WSNs:

- Environmental applications [3, 33, 86, 226]. Some environmental applications of sensor networks include tracking the movements of birds, small animals, and insects; monitoring environmental conditions that affect crops and livestock; irrigation; macro-instruments for large-scale Earth monitoring and planetary exploration; chemical/biological detection; precision agriculture; biological, Earth, and environmental monitoring in marine, soil, and atmospheric contexts; forest fire detection; meteorological or geophysical research; bio-complexity mapping of the environment; and pollution study.
- Health applications [62]. Some of the health applications for sensor networks are providing interfaces for the disabled; integrated patient monitoring; diagnostics; drug administration in hospitals; constant monitoring of human physiological data; exact micro-drug release and non-invasive surgery; telemonitoring of elderly people.
- Other commercial applications [3, 86, 182]. Some of the commercial applications are monitoring material fatigue; building virtual keyboards; managing inventory; monitoring product quality; constructing smart office spaces; environmental control in office buildings; robot control and guidance in automatic manufacturing environments; interactive toys; interactive museums; factory process control and automation; monitoring disaster areas; smart structures with sensor nodes

embedded inside; machine diagnosis; transportation; factory instrumentation; local control of actuators; detecting and monitoring car thefts; vehicle tracking and detection.

- Military applications [3]. Since WSNs are fault-tolerant, self-organized, miniaturized, low cost, and can be easily deployed (for instance, spread by a helicopter), they can be considered as a valuable resource for the military. Some of the military applications of sensor networks could be: Monitoring battlefield resources; battlefield surveillance; nuclear, biological and chemical (NBC) agents detection and reconnaissance; tactical communications.

1.1.2 Enabling Technologies

In this section we briefly review the main enabling technologies for WSN [174]:

- Sensors Components. A sensor node is made up of four basic components as shown in Fig. 1.1: A sensing unit, a processing unit, a transceiver unit and a power unit. They may also have additional application-dependent components such as a location finding system, a power generator and a mobilizer. Sensing units are usually composed of two subunits: Sensors and Analog to Digital Converters (ADCs). The analog signals produced by the sensors based on the observed phenomenon are converted into digital signals by the ADC, and then fed into the processing

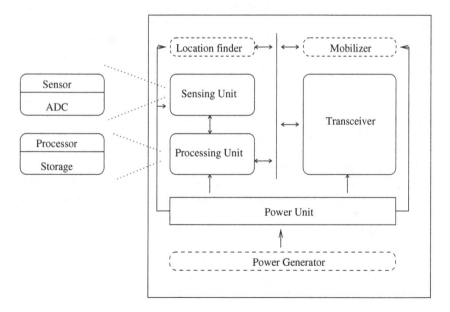

Fig. 1.1 Logical components of a sensor unit

unit. The processing unit, which is generally associated with a small storage unit, manages the procedures that make the sensor node collaborate with the other nodes to carry out the assigned sensing tasks. A transceiver unit connects the node to the network. One of the most important components of a sensor node is the power unit. There can be also other subunits, which are application-dependent.

Many of the sensor network routing techniques and sensing tasks require the knowledge of the location with high accuracy. Thus, in some application a sensor node can need a location finding system. In some cases it is assumed that each sensor node (or some of them) will have a Global Positioning System (GPS) unit. A mobilizer may sometimes be needed to move sensor nodes when it is required to carry out the assigned tasks. All of these subunits may need to fit into a matchbox-sized module [113, 125, 173]. Power is also a scarce resource due to the size limitations. Power units may be supported by a power scavenging unit such as solar cells, or equipped with the capability of transforming vibration into energy [39, 189]. For instance, the total stored energy in a smart dust mote is on the order of 1 J [113, 180]. Though the higher computational powers are being made available in smaller and smaller processors, processing and memory units of sensor nodes are still scarce resources. For instance, the processing unit of a smart dust mote prototype is a 4 MHz Atmel AVR8535 micro-controller with 8 KB instruction flash memory, 512 bytes RAM and 512 bytes EEPROM [113, 173].

- Communication Technology. One of the communication technologies promising for WSN is the Bluetooth technology. This standard provides ad hoc configuration of master/slave piconets including eight active units at most. It supports spontaneous connections between devices. Bluetooth allows data transfers between units over distances of nominally up to 10 m. Bluetooth was originally conceived as a cable replacement technology and may serve well in that application domain. With data-rates of up to 1 Mbps, Bluetooth also offers more than enough bandwidth for our purpose. However, scenarios involving a large number of low-power devices using ad hoc networking still face a number of obstacles when using Bluetooth as their communication technology [130].

  Another technology of interest is the one coming from the ZigBee working group [120], also known as PURLnet, RF-Lite, Firefly, and HomeRF Lite. It is yet another initiative to develop a standard for a small range wireless network. Its projected application areas include computer peripherals, toys, home automation (light, fire alarms, etc.) and remote controls. Depending on the application area, the nodes shall operate between a month and two years on AA-batteries. The nodes are aimed to cost less than 5 USD, operate in the 2.4 GHz ISM band and provide a data rate between 115 and 10 kbps per node. The maximum number of nodes is 255. A recent standardization process has been initiated by the IEEE 802.15 working group [1], which tries to define a Personal Area Network (PAN) standard. Its first incarnation (802.15.1) is to be based on Bluetooth and should improve and extend the existing specification. 802.15.3 aims for high data rates of 20 Mbps or more. Recently ZigBee was standardized as IEEE 802.15.4. The accepted frequency bands are 868/915 MHz and 2.45 GHz. The communication range is 75 m. The communication can only be between one master and 250 slaves.

- Operating System. Operating systems in embedded wireless communication must increasingly satisfy a tight set of constraints, such as power and real time performance, on heterogeneous software and hardware architectures. In this domain, it is well understood that traditional general-purpose operating systems are not efficient or in many cases not sufficient. Among the most emerging solutions for sensor node operating systems we have TinyOS [113]. It is a component-based runtime environment designed to provide support for deeply embedded systems, which require concurrency intensive operations while constrained by minimal hardware resources [113]. The main advantage of the TinyOS is its very small code size, and thus well suited for sensor nodes equipped with a minimum level of hardware. The core scheduler of the operating system fits into only 178 bytes of memory, propagates events in the time it takes to copy 1.25 bytes of memory, context-switches in the time it takes to copy 6 bytes of memory and supports two-level scheduling. The network for which it was designed is a multi-hop network made up of static nodes. It is a centralized system; a base station is conducting the data acquisition and the communication between the nodes. TinyOS is a power-aware operating system. Unused clock cycles are spent in the sleep mode. The software components are written so that they perform their function and go to sleep mode. If any data arrives, the event is signaled to the appropriate component. Software components can also ask the task scheduler to perform tasks (blocks of code that run to completion). It can be the starting point for another more general operating system that assumes: (i) there is no base station, and (ii) the nodes are mobile. Other alternative operating system solutions can include: Microsoft Windows CE [116], a scaled down operating system designed specifically for what Microsoft terms "information appliances"; Palm OS [117]; and Redhat eCos [118], an open source, real-time operating system that provides the basic runtime infrastructure with memory footprints in a limited storage space.

1.1.3 Constraints

A wireless sensor network is a special network which has many constraints compared to a traditional computer network. Due to these constraints it is difficult to directly employ the existing security approaches to the area of wireless sensor networks. Therefore, in order to develop useful security mechanisms while borrowing the ideas from the current security techniques, it is necessary to know and understand these constraints first [30].

1.1.3.1 Very Limited Resources

All security approaches require a certain amount of resources for the implementation, including data memory, code space, and energy to power the sensor. However, currently these resources are very limited in a tiny wireless sensor:

- One common sensor type (TelosB) has a 16-bit, 8 MHz RISC CPU with only 10 K RAM, 48 K program memory, and 1024 K flash storage [119]. With such a limitation, the software built for the sensor must also be quite small. The total code space of TinyOS, the de-facto standard operating system for wireless sensors, is approximately 4 K [113], and the core scheduler occupies only 178 bytes. Therefore, the code size for the all security related code must also be small.
- Power Limitation. Energy is the biggest constraint to wireless sensor capabilities. We assume that once sensor nodes are deployed in a sensor network, they cannot be easily replaced (high operating cost) or recharged (high cost of sensors). Therefore, the battery charge taken with them to the field must be conserved to extend the life of the individual sensor node and the entire sensor network. When implementing a cryptographic function or protocol within a sensor node, the energy impact of the added security code must be considered. When adding security to a sensor node, we are interested in the impact that security has on the lifespan of a sensor (i.e., its battery life). The extra power consumed by sensor nodes due to security is related to the processing required for security functions (e.g., encryption, decryption, signing data, verifying signatures), the energy required to transmit the security-related data or overhead (e.g., initialization vectors needed for encryption/decryption), and the energy required to store security parameters in a secure manner (e.g., cryptographic key storage).

1.1.3.2 Unreliable Communication

Unreliable communication is another threat to sensor security. The security of the network relies heavily on a defined protocol, which in turn depends on communication:

- Unreliable Transfer. Normally the packet-based routing of the sensor network is connectionless and thus inherently unreliable. Packets may get damaged due to channel errors or dropped at highly congested nodes. The result is lost or missing packets. Furthermore, the unreliable wireless communication channel also results in damaged packets. An higher channel error rate also forces the software developer to devote resources to error handling. More importantly, if the protocol lacks the appropriate error handling it is possible to lose critical security packets. This may include, for example, a cryptographic key.
- Conflicts. Even if the channel is reliable, the communication may still be unreliable. This is due to the broadcast nature of the wireless sensor network. If packets meet in the middle of transfer, conflicts will occur and the transfer itself will fail. In a crowded (high density) sensor network, this can be a major problem. More details about the effect of wireless communication can be found at [3].
- Latency. Multi-hop routing, network congestion, and node processing can lead to greater latency in the network, thus making it difficult to achieve synchronization among sensor nodes. The synchronization issues can be critical to sensor security where the security mechanism relies on critical event reports and cryptographic

key distribution. Real-time communications issues in wireless sensor networks are discussed in [208].

1.1.3.3 Unattended Operation

Depending on the function of the particular sensor network, the sensor nodes may be left unattended for long periods of time. There are three main caveats to unattended sensor nodes:

- Exposure to Physical Attacks. The sensor may be deployed in an environment open to adversaries, bad weather, and so on. The likelihood that a sensor suffers a physical attack in such an environment is therefore much higher than the typical PCs, which are located in a secure place and mainly face attacks from a network.
- Remote Management. Remote management of a sensor network makes it virtually impossible to detect physical tampering (i.e., through tamper-proof seals) and physical maintenance issues (e.g., battery replacement). Perhaps the most extreme example of this is a sensor node used for remote reconnaissance missions behind enemy lines. In such a case, the node may not have any physical contact with friendly forces once deployed.
- No Central Management Point. A sensor network can be a distributed network without a central management point. This will increase the vitality of the sensor network. However, if designed incorrectly, it will make the network organization difficult, inefficient, and fragile.

Perhaps most importantly, the longer a sensor is left unattended the more an adversary is likely to compromise the node.

1.2 Security Issues in Wireless Sensor Networks

In this section we report an overview of the security issues in Wireless Sensor Networks [175, 221]. In particular, Sect. 1.2.1 presents the typical security requirements under the context of WSNs. Section 1.2.2 describes the main attacks in WSNs while Sect. 1.2.3 presents some of the main countermeasures. The following chapters of this book provide a deeper discussion of a subset of the aspects presented in this section. Chapter 2 presents a new mechanism for the requirements of authentication and confidentiality. Chapters 3 and 4 address two specific attacks: Node capture and node cloning, respectively. Finally, Chaps. 5 and 6 address two different security aspects of data aggregation: Resilience to malicious node presence and node privacy, respectively.

1.2.1 Security Requirements and Related Issues

A sensor network is a special type of network. It shares some commonalities with a typical computer network, but also poses unique requirements of its own as discussed in Sect. 1.1. Therefore, we can think of the requirements of a wireless sensor network as encompassing both the typical network demands and the unique necessities suited solely to wireless sensor networks.

1.2.1.1 Data Confidentiality

Data confidentiality is the most important issue in network security. Every network with any security focus will typically address this problem first. In sensor networks, the confidentiality relates to the following [30, 173]:

- A sensor network should not leak sensor readings to its neighbours.
- In many applications nodes communicate highly sensitive data, e.g., key distribution, therefore it is extremely important to build a secure channel in a wireless sensor network.
- Public sensor information, such as sensor identities and public keys, should also be encrypted to some extent to protect against traffic analysis attacks.

The standard approach for keeping sensitive data secret is to encrypt the data with a secret key that only intended receivers possess, thus achieving confidentiality.

1.2.1.2 Authentication

An adversary is not just limited to modify the data packet. It can change the whole packet stream by injecting additional packets. So the receiver needs to ensure that the data used in any decision-making process originates from the correct source. On the other hand, when constructing the sensor network, authentication is necessary for many administrative tasks (e.g., network reprogramming or controlling sensor node duty cycle). From the above, we can see that message authentication is important for many applications in sensor networks. Informally, data authentication allows a receiver to verify that the data is really sent by the claimed sender. In the case of two-party communication, data authentication can be achieved through a purely symmetric mechanism: The sender and the receiver share a secret key to compute the Message Authentication Code (MAC) of all communicated data. Adrian Perrig et al. propose a key-chain distribution system for their μTESLA secure broadcast protocol [173]. The basic idea of the μTESLA system is to achieve asymmetric cryptography by delaying the disclosure of the symmetric keys. One limitation of μTESLA is that some initial information must be unicast to each sensor node before authentication of broadcast messages can begin. Liu and Ning [143, 145] propose an enhancement to the μTESLA system that uses broadcasting of the key chain commitments rather than μTESLA's unicasting technique.

1.2.1.3 Data Integrity

With the implementation of confidentiality, an adversary may be unable to steal information. However, this does not mean the data is safe. The adversary can change the data, so as to send the sensor network into disarray. For example, a malicious node may add some fragments or manipulate the data within a packet. This new packet can then be sent to the original receiver. Data loss or damage can even occur without the presence of a malicious node due to the harsh communication environment. Thus, data integrity ensures that any received data has not been altered in transit.

1.2.1.4 Data Freshness

Even if confidentiality and data integrity are assured, we also need to ensure the freshness of each message. Informally, data freshness suggests that the data is recent, and it ensures that no old messages have been replayed. This requirement is especially important when there are shared-key strategies employed in the design. Typically shared keys need to be changed over time. However, it takes time for new shared keys to be propagated to the entire network. In this case, it is easy for the adversary to use a replay attack. Also, it is easy to disrupt the normal work of the sensor, if the sensor is unaware of the new key change time. To solve this problem a nonce, or another time-related counter, can be added into the packet to ensure data freshness.

1.2.1.5 Availability

Adjusting the traditional encryption algorithms to fit within the wireless sensor network will introduce some extra costs. Some approaches choose to modify the code to reuse as much code as possible. Some approaches try to make use of additional communication to achieve the same goal. What is more, some approaches force strict limitations on the data access, or propose an unsuitable scheme (such as a central point scheme) in order to simplify the algorithm. But all these approaches weaken the availability of a sensor and sensor network for the following reasons:

- Additional computation consumes additional energy. If no more energy exists, the data will no longer be available.
- Additional communication also consumes more energy. What is more, as communication increases so it does the chance of incurring a communication conflict.
- A single point of failure will be introduced if using the central point scheme. This greatly threatens the availability of the network.

The requirement of security not only affects the operation of the network, but also is highly important in maintaining the availability of the whole network.

1.2.1.6 Self-Organization

A wireless sensor network is typically an ad hoc network, which requires every sensor node to be independent and flexible enough to be self-organizing and self-healing according to different situations. There is no fixed infrastructure available for the purpose of network management in a sensor network. This inherent feature brings a great challenge to wireless sensor network security as well. For example, the dynamics of the whole network inhibits the idea of pre-installation of a shared key between the base station and all sensors [85]. Several random key pre-distribution schemes have been proposed in the context of symmetric encryption techniques [37, 85, 122, 144]. In the context of applying public-key cryptography techniques in sensor networks, an efficient mechanism for public-key distribution is necessary as well. In the same way that distributed sensor networks must self-organize to support multihop routing, they must also self-organize to conduct key management and building trust relation among sensors. If self-organization is lacking in a sensor network, the damage resulting from an attack or even the hazardous environment may be devastating.

1.2.1.7 Time Synchronization

Most sensor network applications rely on some form of time synchronization. In order to conserve power, an individual sensor's radio may be turned off for periods of time. Furthermore, sensors may wish to compute the end-to-end delay of a packet as it travels between two pairwise sensors. A more collaborative sensor network may require group synchronization for tracking applications, etc. In [92], the authors propose a set of secure synchronization protocols for sender-receiver (pairwise), multihop sender-receiver (for use when the pair of nodes are not within single-hop range), and group synchronization.

1.2.1.8 Secure Localization

Often, the utility of a sensor network will rely on its ability to accurately and automatically locate each sensor in the network. A sensor network designed to locate faults will need accurate location information in order to pinpoint the location of a fault. Unfortunately, an attacker can easily manipulate non-secured location information by reporting false signal strengths, replaying signals, etc. A technique called Verifiable Multilateration (VM) is described in [216]. In multilateration, a device's position is accurately computed from a series of known reference points. In [216], authenticated ranging and distance bounding are used to ensure accurate location of a node. Because of distance bounding, an attacking node can only increase its claimed distance from a reference point. However, to ensure location consistency, an attacking node would also have to prove that its distance from another reference point is shorter [216]. Since it cannot do this, a node manipulating the localization protocol can be found. For large sensor networks, the SPINE (Secure Positioning for sensor NEtworks) algorithm is used. It is a three-phase algorithm based upon verifiable multilateration [216]. In [139], SeRLoc (Secure Range-Independent Localization) is described. Its novelty is its decentralized, range-independent nature. SeRLoc uses locators that transmit beacon information. It is assumed that the locators are trusted and cannot be compromised. Furthermore, each locator is assumed to know its own location. A sensor computes its location by listening for the beacon information sent by each locator. The beacons include the locator's location. Using all of the beacons that a sensor node detects, a node computes an approximate location based on the coordinates of the locators. Using a majority vote scheme, the sensor then computes an overlapping antenna region. The final computed location is the "center of gravity" of the overlapping antenna region [139]. All beacons transmitted by the locators are encrypted with a shared global symmetric key that is pre-loaded to the sensor prior to deployment. Each sensor also shares a unique symmetric key with each locator. This key is also pre-loaded on each sensor.

1.2.2 Attacks

Sensor networks are particularly vulnerable to several types of attacks. Attacks can be performed in a variety of ways ranging from denial of service to traffic analysis, privacy violation, physical attacks, and so on. In this section, we review some of the most common and studied type of attacks.

1.2.2.1 Denial of Service Attacks

Wood and Stankovic define the denial of service attack as "any event that diminishes or eliminates a network's capacity to perform its expected function" [230]. Denial of service attacks assume a particular importance in wireless sensor networks, where is not possible to afford the computational overhead necessary in implementing many of the typical defensive strategies of traditional computing.

A standard attack on wireless sensor networks is simply to jam a node or set of nodes. Jamming, in this case, is simply the transmission of a radio signal that interferes with the radio frequencies being used by the sensor network [230]. The jamming of a network can come in two forms: Constant jamming and intermittent jamming. Constant jamming involves the complete jamming of the entire network. No messages are able to be sent or received. If the jamming is only intermittent, then nodes are able to exchange messages periodically, but not consistently. This too can have a detrimental impact on the sensor network as the messages being exchanged between nodes may be time sensitive [230].Attacks can also be made on the link layer itself. One possibility is that an attacker may simply intentionally violate the communication protocol, e.g., ZigBee [120] or IEEE 801.11b (Wi-Fi) protocol, and continually transmit messages in an attempt to generate collisions. Such collisions would require the retransmission of any packet affected by the collision. Using this technique it would be possible for an attacker to simply deplete a sensor node's power supply by forcing too many retransmissions. At the routing layer, a node may take advantage of a multihop network by simply refusing to route messages. This could be done intermittently or constantly with the net result being that any neighbour who routes through the malicious node will be unable to exchange messages with, at least, part of the network. Extensions to this technique including intentionally routing messages to incorrect nodes (misdirection) [230]. The transport layer is also susceptible to attack, as in the case of flooding. Flooding can be as simple as sending many connection requests to a susceptible node. In this case, resources must be allocated to handle the connection request. Eventually a node's resources will be exhausted, thus rendering the node useless. Finally, a denial of service attack can be performed against the specific application level protocol. An example can be the disruption of a data aggregation protocol, avoiding the collector node to collect any aggregated data. This specific attack, together with a similar one (where the aim of the attacker is to let the collecting node to accept a false aggregated value), are addressed in Chap. 5.

1.2.2.2 The Sybil Attack

The sybil attack can be defined as a "malicious device illegitimately taking on multiple identities" [159]. It was originally described as an attack able to defeat the redundancy mechanisms of distributed data storage systems in peer-to-peer networks [79]. In addition, the sybil attack is also effective against routing algorithms, data aggregation, voting, fair resource allocation and foiling misbehavior detection. Regardless of the target (voting, routing, aggregation) all of the attacking techniques involve the use of multiple identities. For instance, in a voting scheme, the sybil attack might utilize multiple identities to generate additional "votes". Similarly, to attack the routing protocol, the sybil attack would rely on a malicious node taking on the identity of multiple nodes, and thus routing multiple paths through a single malicious node.

1.2.2.3 Traffic Analysis Attacks

Wireless sensor networks are typically composed of many low-power sensors communicating with a few relatively robust and powerful base stations. It is not unusual, therefore, for data to be gathered by the individual nodes where it is ultimately routed to the base station. Often, for an adversary to effectively render the network useless, the attacker can simply disable the base station. To make matters worse, Deng et al.

demonstrate two attacks that can identify the base station in a network (with high probability) without even understanding the contents of the packets (if the packets are themselves encrypted) [69].

A rate monitoring attack simply makes use of the idea that nodes closest to the base station tend to forward more packets than those farther away from the base station. An attacker need only monitor which nodes are sending packets and follow those nodes that are sending the most packets. In a time correlation attack, an adversary simply generates events and monitors to whom a node sends its packets. To generate an event, the adversary could simply generate a physical event that would be monitored by the sensor(s) in the area (turning on a light, for instance) [69].

1.2.2.4 Node Replication Attacks

A node replication attack is similar to the sybil attack. Instead of having a node using more identities (as for the sybil attack), the attacker here seeks to add a node to an existing sensor network by copying (replicating) a node's memory into the memory of a newly created node [169]. The power of this attack is that, cloning both the ID of the original nodes and the cryptographic material used to prove the honesty of the corresponding ID, it is hard to detect that we are facing with a bogus node. A node replicated in this fashion can severely disrupt a sensor network's performance: Packets can be corrupted or even misrouted. This can result in a disconnected network, false sensor readings, etc. If an attacker can gain physical access to the entire network it can copy cryptographic keys to the replicated sensor and can also insert the replicated node into strategic points in the network [169]. As an example, by inserting the replicated nodes at specific network points, the attacker could easily manipulate a specific segment of the network.

1.2.2.5 Attacks Against Privacy

Sensor network technology promises a vast increase in automatic data collection capabilities through efficient deployment of tiny sensor devices. While these technologies offer great benefits to users, they also exhibit significant potential for abuse. Particularly relevant concerns are privacy problems, since sensor networks provide increased data collection capabilities [105]. Adversaries can use even seemingly innocuous data to derive sensitive information if they know how to correlate multiple sensor inputs. For example, in the famous "panda-hunter problem" [166], the hunter can imply the position of pandas by monitoring the traffic. The main privacy problem, however, is not that sensor networks enable the collection of information. In fact, much information from sensor networks could probably be collected through direct site surveillance. Rather, sensor networks aggravate the privacy problem because they make large volumes of information easily available through remote access. Hence, adversaries need not be physically present to maintain surveillance. They can gather information in a low-risk, anonymous manner. Remote access also allows a single adversary to monitor multiple sites simultaneously [35]. Some of the more common attacks [35, 105] against sensor privacy are:

- Monitor and Eavesdropping. This is the most obvious attack to privacy. By listening to the data, the adversary could easily discover the communication contents. When the traffic conveys the control information about the sensor network configuration, which contains potentially more detailed information than accessible through the location server, the eavesdropping can act effectively against the privacy protection.
- Traffic Analysis. This typically combines with monitoring and eavesdropping. An increase in the number of transmitted packets between certain nodes could signal that a specific sensor has registered activity. Through the analysis on the traffic, some sensors with special roles or activities can be effectively identified.
- Camouflage Adversaries. They can insert their node or compromise the nodes to hide in the sensor network. After that these nodes can masquerade as a normal node to attract the packets, then misroute the packets, e.g. forward the packets to the nodes conducting the privacy analysis.

It is worth noting that, as pointed out in [171], the understanding of privacy in wireless sensor networks is not fully mature.

1.2.2.6 Physical Attacks

Sensor networks typically operate in hostile outdoor environments. In such environments, the tiny size of the sensors, coupled with the unattended and distributed nature of their deployment make them highly susceptible to physical attacks, i.e., threats due to physical node destruction [225]. Unlike many other attacks mentioned above, physical attacks destroy sensors permanently, so the losses are irreversible. For instance, attackers can extract cryptographic secrets, tamper with the associated circuitry, modify programming in the sensors, or replace them with malicious sensors under the control of the attacker [106]. Hartung et al. showed that standard sensor nodes, such as the MICA2 motes, can be compromised in less than one minute [109]. While these results are not surprising given that the MICA2 lacks tamper resistant hardware protection, they provide a cautionary note about the speed of a well-trained attacker. If an adversary compromises sensor node, then the code inside the physical node may be modified.

1.2.3 Defensive Measures

In this section, we describe some of the main security defensive measures ranging from key establishment, which lays the foundation for different security aspects, to more specific measures such as those for the routing or defending against attacks on sensor privacy.

1.2.3.1 Key Establishment

One security aspect that deserves a great attention in wireless sensor networks is the area of key management. Wireless sensor networks are unique (among other embedded wireless networks) in this aspect due to their size, mobility and computational/power constraints. This, coupled also with the typical operational constraints of WSNs, makes secure key management an absolute necessity in most wireless sensor network designs.

In traditional network, key establishment is done using public-key protocols such as the Diffie–Hellman protocol [77]. Most of the traditional techniques, however, are unsuitable in low power devices such as wireless sensor networks. The problem with asymmetric cryptography, in a wireless sensor network, is that it is typically too computationally intensive for the individual nodes in a sensor network. While this is true in the general case, different researchers show that it is feasible with the right selection of algorithms [95, 107, 149, 227]. Two of the major techniques used to implement public-key cryptosystems are RSA and Elliptic Curve Cryptography (ECC) [200]. In [227], Watro et al. show that portions of the RSA cryptosystem can be successfully applied to actual wireless sensors, specifically the UC Berkeley MICA2 motes [113]. In [149], Malan et al. demonstrate a working implementation of Diffie–Hellman based on the Elliptic Curve Discrete Logarithm Problem. In [238], Zanin et al. propose a distributed and efficient signature protocol for ad hoc network with tight security and architectural constraints, as WSNs are.

Despite the actual possibility of implementation of asymmetric cryptography on sensor nodes, symmetric cryptography is the typical choice for applications that cannot afford the computational complexity of asymmetric cryptography. Symmetric schemes utilize a single shared key known only between the two communicating hosts. Traditional examples of symmetric cryptographic algorithm are DES (Data Encryption Standard), 3DES (Triple DES), RC5, and AES [200]. An analysis of the various ciphers is presented in [136]. One major shortcoming of symmetric cryptography is the key exchange problem: Two communicating hosts must somehow know the shared key before they can communicate securely. As extreme cases we have the following:

- All the nodes shares the same master symmetric key. In this case each pair of nodes can communicate but if a node is compromised all the communications are also compromised.
- Each pair of nodes shares a different symmetric key. As a result, each node has to store a huge amount of keys.

To mitigate the problems related to these extreme cases different variant of random key pre-distribution schemes have been proposed [37, 85, 122, 144].

Other authentication techniques for WSNs that make no use of pre-established or pre-certified keys have also been proposed. As an example, in [219] the authors propose an authentication mechanism based on the concept of integrity regions: The entity proximity is verified through time-of-arrival ranging techniques.

The key establishment problem is further discussed in Chap. 2, where a new pairwise key establishment scheme, the ECCE Protocol, is also presented.

1.2.3.2 Defending Against DoS Attacks

Since denial of service attacks are common and highly effective in WSN, defenses must be available to combat them.

One strategy in defending against the classic jamming attack is to identify the jammed part of the sensor network and effectively route around the unavailable portion. Wood and Stankovic [230] describe a two-phase approach where the nodes along the perimeter of the jammed region report their status to their neighbours who then collaboratively define the jammed region and simply route around it. To handle jamming at the MAC layer, nodes might utilize a MAC admission control that is rate limiting. This would allow the network to ignore those requests designed to exhaust the power reserves of a node. This, however, is not fool-proof as the network must be able to handle any legitimately large traffic volume. Overcoming rogue sensors that intentionally misroute messages can be done at the cost of redundancy. In this case, a sending node can send the message along multiple paths in an effort to increase the likelihood that the message will ultimately arrive at its destination. This has the advantage of effectively dealing with nodes that may not be malicious, but rather may have simply failed as it does not rely on a single node to route its messages. To overcome the transport layer flooding denial of service attack, Aura et al. [13] suggest using the client puzzles posed by Juels and Brainard [13] in an effort to discern a node's commitment to making the connection by utilizing some of their own resources. Aura et al. advocate that a server should force a client to commit its own resources first. Further, they suggest that a server should always force a client to commit more resources up front than the server. This strategy would likely be effective as long as the client has computational resources comparable to those of the server.

1.2.3.3 Secure Broadcasting and Multicasting

In wireless sensor networks, a great deal of the security derives from ensuring that only members of the broadcast or multicast group possess the required keys in order to decrypt the broadcast or multicast messages. Because of this, most of the key-management solutions are applicable. Here we address those schemes that have been specifically designed to support broadcasting and multicasting in wireless sensor networks.

In traditional network different key management schemes have been devised: Centralized group key management protocols, decentralized management protocols, and distributed management protocols [69]. In order to efficiently distribute keys, one well known technique is to use a logical key tree. Such a technique falls into the centralized group key management protocols. This technique has been extended to wireless sensor networks in [137, 138, 176]. While centralized solutions are often not ideal, in the case of wireless sensor networks a centralized solution offers some utility. Such a technique allows a more powerful base station to offload some of the computations from the less powerful sensor nodes.

Di Pietro et al. describe a directed diffusion based multicast technique for use in wireless sensor networks that also takes advantage of a logical key hierarchy [176]. Directed diffusion is a data-centric, energy efficient dissemination technique that has been designed for use in wireless sensor networks [125]. Using the above mentioned directed diffusion technique, Di Pietro et al. enhance the logical key hierarchy to create a directed diffusion based logical key hierarchy.

Kaya et al. discuss the problem of multicast group management in [131]. In this case, nodes are grouped based on locality and attach to a security tree. However, they assume that nodes within the mobile network are somewhat more powerful than a traditional sensor in a wireless sensor network.

Lazos and Poovendran describe a tree based key distribution scheme that is similar to [176]. They suggest a routing-aware based tree where the leaf nodes are assigned keys based on all relay nodes above them. They argue that their technique, which takes advantage of routing information, is more energy efficient than routing schemes that arbitrarily arrange nodes into the routing tree. They propose a greedy routing-aware key distribution algorithm [137]. In [138], Lazos and Poovendran use a similar technique to [137], but instead use geographic location information (e.g., GPS) rather than routing information. In this case, however, nodes (with the help of the geographic location system) are grouped into clusters with the observation that nodes within a cluster will be able to reach one another with a single broadcast. Using the cluster information, a key hierarchy is constructed as in [137].

1.2.3.4 Defending Against Attacks on Routing Protocols

As for the routing in wireless sensor networks, most current research has focused primarily on providing the most energy efficient routing. However, there is a great need for both secure and energy efficient routing protocols in wireless sensor networks against attacks such as the sinkhole, wormhole and sybil attacks [121, 128, 159]. A discussion on many of the attacks on routing protocols is given in [128].

In general, packet routing algorithms are used to exchange messages with sensor nodes that are outside of a particular radio range. This is different than sensors that are within radio range where packets can be transmitted using a single hop. In such single hop networks security is still a concern, but is more accurately addressed through secure broadcasting and multicasting. The first packet routing algorithm is based on node identifiers similar to traditional routing. In this case, each sensor is identified by an address and routing to/from the sensor is based on the address. This is generally considered inefficient in sensor networks, where nodes are expected to be addressed by their location, rather than their identifier. As a consequence of the distaste of routing based on node identifiers, geographic routing protocols have been introduced [23, 129]. One common routing protocol, GPSR [129] allows nodes to send a packet to a region, rather than a particular node. Such a routing protocol lends itself nicely to the concept of data-centric networks. Security specific to this type of network is discussed in [212].

As for techniques for securing the routing protocols, Deng, Han, and Mishra describe an intrusion tolerant routing protocol, INSENS, that is designed to limit the scope of an intruder's destruction and route despite network intrusion without having to identify the intruder [70]. Tanachaiwiwat, et al. present a novel technique named TRANS (Trust Routing for Location Aware Sensor Networks) [212]. The TRANS routing protocol is designed for use in data centric networks. It also makes use of a loose-time synchronization asymmetric cryptographic scheme to ensure message confidentiality. In their implementation, μTESLA is used to ensure message authentication and confidentiality. Using μTESLA, TRANS is able to ensure that a message is sent along a path of trusted nodes while also using location aware routing.

One particular challenge to secure routing in wireless sensor networks is that it is very easy for a single node to disrupt the entire routing protocol by simply disrupting the route discovery process. Papadimitratos and Haas propose a secure route discovery protocol that guarantees, subject to several conditions, that correct topological information will be obtained [167]. The security relies on the MAC and an accumulation of the node identities along the route traversed by a message. In so doing, a source can discover the sensor network topology as each node along the route from source to destination appends its identity to the message. In order to ensure that the message has not been tampered with, a MAC is constructed and can be verified both at the destination and the source (for the return message from the destination). A related problem is the concept of wormholes in a sensor network. A wormhole attack is one in which a malicious node eavesdrops on a packet or series of packets, tunnels them through the sensor network to another malicious node, and then replays the packets. This can be done to misrepresent the distance between the two colluding nodes. It can also be used to more generally disrupt the routing protocol by misleading the neighbour discovery process [128]. Often additional hardware, such as a directional antenna [142], is used to defend against wormhole attacks. This, however, can be cost-prohibitive when it comes to large-scale network deployment. Instead, Wang and Bhargava use a visualization approach to identify wormholes [223]. They first compute a distance estimation between all neighbour sensors, including possible existing wormholes. Using multi-dimensional scaling, they then compute a virtual layout of the sensor network. A surface smoothing strategy is then used to adjust for round-off errors in the multi-dimensional scaling. Finally, the shape of the resulting virtual network is analyzed. If a wormhole exists within the network, the shape of the virtual network will bend and curve towards the offending nodes. Using this strategy the nodes that participate in the wormhole can be identified and removed from the network. If a network does not contain a wormhole, the virtual network will appear flat [223].

1.2.3.5 Defending Against the Sybil Attack

To defend against the sybil attack, the network needs some mechanism to validate that a particular identity is the only identity being held by a given physical node [159]. Newsome et al. describe two methods to validate identities, direct validation and indirect validation. In direct validation a trusted node directly tests whether the joining identity is valid. In indirect validation, another trusted node is allowed to vouch for (or against) the validity of a joining node [159]. Newsome et al. primarily describe direct validation techniques, including a radio resource test. In the radio test, a node assigns each of its neighbours a different channel on which to communicate. The node then randomly chooses a channel and listens. If the node detects a transmission on the channel it is assumed that the node transmitting on the channel is a physical node. Similarly, if the node does not detect a transmission on the specified channel, the node assumes that the identity assigned to the channel is not a physical identity. Another technique to defend against the sybil attack is to use random key pre-distribution techniques. The idea behind this technique is that with a limited number of keys on a keyring, a node that randomly generates identities will not possess enough keys to take on multiple identities and thus will be unable to exchange messages on the network due to the fact that the invalid identity will be unable to encrypt or decrypt messages.

1.2.3.6 Detecting Node Replication Attacks

The detection of node replication attacks becomes a particular challenge in WSN, compared to the traditional network. In particular, the detection should be done in a way as distributed as possible. Researcher only recently started addressing this problem [199, 245]. In 2005 Parno et al. proposed two algorithms for the distributed detection of the clone attack [169]. As further discussed in Chap. 4, dedicated to the clone detection, the Parno et al. solution presents many drawbacks. In Sect. 4.2 we present an overview of the other proposals for solving this problem. Furthermore, in the same chapter we present a new efficient and distributed detection protocol, and we compare it with the other solutions.

1.2.3.7 Defending Against Attacks on Sensor Privacy

Regarding the attacks on privacy mentioned in Sect. 1.2.2, there exist effective techniques to counter many of the attacks levied against a sensor. Here we describe several common techniques [105].

Location information that is too precise can enable the identification of a user, or make the continued tracking of movements feasible. This is a threat to privacy. Anonymity mechanisms depersonalize the data before the data is released, which present an alternative to privacy policy-based access control. Researchers have discussed several approaches using anonymity mechanisms, for example, Gruteser and

Grunwald [103] analyze the feasibility of anonymizing location information for location-based services in an automotive telematics environment; Beresford and Stajano [18] independently evaluate anonymity techniques for an indoor location system based on the Active Bat. Total anonymity is a difficult problem given the lack of knowledge concerning a node's location. Therefore, a tradeoff is required between anonymity and the need for public information when solving the privacy problem. In [104, 105, 181, 205], four main approaches are proposed:

- Decentralize Sensitive Data. The basic idea of this approach is to distribute the sensed location data through a spanning tree, so that no single node holds a complete view of the original data.
- Secure Communication Channel. Using secure communication protocols, such as SPINS [172], the eavesdropping and active attacks can be prevented.
- Change Data Traffic. De-patterning the data transmissions can protect against traffic analysis. For example, inserting some bogus data can intensively change the traffic pattern when needed.
- Node Mobility. Making the sensor movable can be effective in defending privacy, especially the location. For example, the Cricket system [181] is a location-support system for in-building, mobile, location dependent applications. It allows applications running on mobile and static nodes to learn their physical location by using listeners that hear and analyze information from beacons spread throughout the building. Thus the location sensors can be placed on the mobile device as opposed to the building infrastructure, and the location information is not disclosed during the position determination process and the data subject can choose the parties to which the information should be transmitted.

Policy-based approaches are currently a hot approach to address the privacy problem. The access control decisions and authentications are made based on the specifications of the privacy policies. In [155], Molnar and Wagner present the concept of private authentication, and give a general scheme for building private authentication with work logarithmic in the number of tags in (but not limited by) RFID (Radio Frequency IDentification) applications. In the automotive telematics domain, Duri et al. [82] propose a policy-based framework for protecting sensor information, where an in-car computer can act as a trusted agent. Snekkenes [206] presents advanced concepts for specifying policies in the context of a mobile phone network. These concepts enable access control based on criteria such as time of the request, location, speed, and identity of the located object. Myles and colleagues [157] describe an architecture for a centralized location server that controls access from client applications through a set of validator modules that check XML-encoded application privacy policies. Hengartner and Steenkiste [152] point out that access control decisions can be governed by either room or user policies. The room policy specifies who is permitted to find out about the people currently in a room, while the user policy states who is allowed to get location information about another user.

Ozturk et al. propose anti-traffic analysis mechanisms to prevent an outside attacker from tracking the location of a data source, since that information will release the location of sensed objects [166]. The randomized data routing mechanism and phantom traffic generation mechanism are used to disguise the real data traffic, so that it is difficult for an adversary to track the source of data by analyzing network traffic.

Similar mechanisms are also used to disguise an adversary from finding the location of a base station by analyzing network traffic [107]. One key problem for these anti-traffic analysis mechanisms is the energy cost incurred by anonymization. Another strategy used to mask location information from eavesdroppers is presented in [231]. They propose a two way greedy random-walk strategy GROW (Greedy Random Walk).

Threats to node privacy can also come from specific application protocols such as the data aggregation protocol. Threat to privacy in data aggregation is discussed in Chap. 6. In the same chapter a new privacy-preserving aggregation protocol is presented. To the best of our knowledge this is the first aggregation protocol that preserves the privacy of the node (i) against other nodes and (ii) against the base station that collects the aggregated data.

1.2.3.8 Intrusion Detection

We now turn to the area of intrusion detection in wireless sensor networks. It is important to note that in this section we cover intrusion detection as it applies to detecting attacks on the sensor network itself, rather than the popular intrusion detection application being researched for such uses as perimeter monitoring, and so forth. With that in mind, we note that intrusion detection is not necessarily a category into itself, but rather has its place in nearly every aspect of sensor network security. Many secure routing schemes attempt to identify network intruders, and key establishment techniques are used in part to prevent intruders from overhearing network data. Despite the necessity of effective intrusion detection schemes for wireless sensor networks, a good solution has not yet been devised. Of course, this is due largely to the resource constraints present in wireless sensor networks. However, resource constraints are not the only reason. Another problem is that researchers have not yet been able to develop methods of reliably detecting intruders in sensor networks. As such, it is difficult to define characteristics (or signatures) that are specific to a network intrusion as opposed to the normal network traffic that might occur as the result of normal network operations or malfunctions resulting from the environment change.

As for intrusion detection, it has traditionally focused on two major categories: AID (Anomaly based Intrusion Detection), and MID (Misuse Intrusion Detection) [161]. Anomaly based intrusion detection relies on the assumption that intruders will demonstrate abnormal behavior relative to the legitimate nodes. Thus, the object of anomaly based detection is to detect intrusion based on unusual system behavior. In systems based on misuse intrusion detection, the system maintains a database of intrusion signatures. Using these signatures, the system can easily detect intrusions on the network.

Typically a wireless sensor network uses cryptography to secure itself against unauthorized external nodes gaining entry into the network. But cryptography can only protect the network against the external nodes and does little to thwart malicious nodes that already possess one or more keys. Brutch and Ko classify IDS (Intrusion Detection Systems) into two categories: Host-based and network-based. They further classify intrusion detection schemes into those that are signature based, anomaly based, and specification based [27]. Simply put, a host based IDS system operates on operating systems audit trails, system call audit trails, logs, and so on. A network based IDS, on the other hand, operates entirely on packets that have been captured from the network. A signature based IDS simply monitors the network for specific pre-determined signatures that are indicative of an intrusion. In an anomaly based scheme, a standard behavior is defined and any deviation from that behavior triggers the intrusion detection system. Finally, a specification based scheme defines a set of constraints that are indicative of a program's or protocol's correct operation. Brutch and Ko describe a series of attacks against several aspects of a wireless sensor network and also introduce three architectures for intrusion detection in wireless sensor networks. The first is termed the stand-alone architecture. In this case, as its name implies, each node functions as an independent intrusion detection system and is responsible for detecting attacks directed toward itself. Nodes do not cooperate in any way. The second architecture is the distributed and cooperative architecture. In this case, an intrusion detection agent still resides on each node (as in the case of the stand-alone architecture) and nodes are still responsible for detecting attacks against themselves (local attacks), but also cooperate to share information in order to detect global intrusion attempts. The third technique proposed by Brutch and Ko is called the hierarchical architecture. These architectures are suitable for multi-layered wireless sensor networks. In this case, Brutch and Ko describe a multi-layered network as one in which the network is divided into clusters with cluster-head nodes responsible for routing within the cluster. The multi-layered network is used primarily for event correlation. Albers et al. describe an intrusion detection architecture based on the implementation of a LIDS (Local Intrusion Detection System) at each node [4]. In order to extend each node's "vision" of the network, Albers suggests that the LIDS existing within the network should collaborate with one another. All LIDS within the network will exchange two types of data, security data and intrusion alerts. The security data is simply used to exchange information with other network hosts. The intrusion alerts, however, are used to inform other LIDS of a locally detected intrusion [4]. Albers et al. propose to use SNMP auditing as the audit source for each LIDS. Rather than simply sending the SNMP messages over an unreliable UDP connection, it is suggested that mobile agents will be responsible for message transporting. In order to detect an intrusion, Albers suggests using either misuse or anomaly detection. When a LIDS detects an intrusion, it should communicate this intrusion to other LIDS on the network. Possible responses include forcing the potential intruder to re-authenticate, or to simply ignore the suspicious node when performing cooperative actions [4]. Although this approach can not be applied to wireless sensor network directly, it is an interesting idea that explores the local information only, which is the key to any intrusion detection techniques in sensor network [86]. In summary, we envision that the intrusion detection in wireless sensors remains an open problem, and more study is needed. Taking the pre-deployment information, such as sensing data distribution, into consideration is a possible direction.

1.2.3.9 Secure Data Aggregation

Due to the computational constraints of sensors, a single node is typically responsible for only a small part of the overall data. Because of this, a query of the wireless sensor network is likely to return a great deal of raw data, much of which is not of interest to the individual performing the query. Furthermore, each node sending its data independently to the collecting node would result in a huge energy consumption. Thus, it is advantageous for the raw data to first be processed so that more meaningful data can be gleaned from the network while saving nodes' energy.

Data aggregation techniques come in help. However, such techniques are particularly vulnerable to attacks as a single node is used to aggregate multiple data. Because of this, secure information aggregation techniques are needed in wireless sensor networks where one or more nodes may be malicious.

Current secure data aggregation protocols are discussed in Chap. 5. In the same chapter, we propose two new algorithms for secure Median computation.

1.2.3.10 Defending Against Physical Attacks

Physical attacks, as we argued in Sect. 1.2.2, pose a great threat to wireless sensor networks, because of its unattended feature and limited resources. Sensor nodes may be equipped with physical hardware to enhance protection against various attacks. For example, to protect against tampering with the sensors, one defense involves tamper-proofing the node's physical package [230]. In [9, 10, 132], the authors focus on building tamper-resistant hardware in order to make the actual data and memory contents on the sensor chip inaccessible to attack. Another way is to employ special software and hardware outside the sensor to detect physical tampering. As the price of the hardware itself gets cheaper, tamper-resistant hardware may become more appropriate in a variety of sensor network deployments. One possible approach to protect the sensors from physical attacks is self-termination. The basic idea is that the sensor kills itself, including destroy all data and keys, when it senses a possible attack. This is particularly feasible in the large scale wireless sensor network which has enough redundancy of information, and the cost of a sensor is much cheaper than the lost of being broken (attacked). The key of this approach is detecting the physical attack. A simple solution is periodically conducting neighborhood checking in static deployment. For mobile sensor networks, this is still an open problem. In [9, 10, 132], the authors describe techniques for extracting protected software and data from smart-card processors. This includes manual microprobing, laser cutting, focused ion-beam manipulation, glitch attacks, and power analysis, most of which are also possible physical attacks on the sensor. Based on an analysis of these attacks,

Andersen et al. give examples of low-cost protection countermeasures that make such attacks considerably more difficult [10].

For the deployment of components outside the sensor Sastry et al. [198] introduce the concept of secure location verification and propose a secure localization scheme, the ECHO protocol, to make sure the location claims are legitimate. In their work, the security rests on physical properties of sound and RF signal propagation. An adversary cannot cheat and claim a shorter distance by starting the ultra-sound response early, because it will not have the nonce. Hu et al. [142] introduce directional antennas to defend against wormhole attacks. In [224] the authors study the modeling and defense of sensor networks against Search-based Physical Attacks. They define a search-based physical attack model, where the attacker walks through the sensor network using signal detecting equipment to locate active sensors, and then destroys them. In a prior work, they have identified and modeled blind physical attacks [223]. The defense algorithm is executed by individual sensors in two phases: In the first phase, sensors detect the attacker and send out attack notification messages to other sensors; in the second phase, the recipient sensors of the notification message schedule their states to switch. A mechanism named SWATT to verify whether the memory of a sensor node has been changed [200] is proposed by Seshadri et al.

The physical capture of a node also implies different consequences, e.g. threatening the communication confidentiality or the data survivability [177]. In Chap. 3, the problem of the physical capture of a node is discussed; A new approach for the node capture detection exploiting the node mobility is also proposed.

1.2.3.11   Trust Management

Trust is an old but important issue in any networked environment, whether social networking or computer networking. Trust can solve some problems beyond the power of the traditional cryptographic security. For example, judging the quality of the sensor nodes and the quality of their services, and providing the corresponding access control, e.g., does the data aggregator perform the aggregation correctly? Does the forwarder send out the packet in a timely fashion? These questions are important, but difficult, if not impossible, to answer using existing security mechanisms. We argue that trust management is the key to build trusted, dependable wireless sensor network applications. The trust issue is emerging as sensor networks thrive. However, it is not easy to build a good trust model within a sensor network given the resource limits. Furthermore, in order to keep the sensor nodes independent, we should not assume there is a trust among sensors in advance.

According to the small world principle in the context of social networks and peer-to-peer computing [162], one can employ a path-finder to find paths from a source node to a designated target node efficiently. Based on this observation, Zhu et al. [246] provide a practical approach to compute trust in wireless networks. They consider individual mobile devices as a node of a delegation graph G. They map a delegation path from the source node, S, to the target node, T, into an edge in the correspondent transitive closure of the graph G, from which the trust value is computed. In this approach, an undirected transitive signature scheme is used within the authenticated transitive graphs. In [233], a trust evaluation based security solution is proposed to provide effective security decisions on data protection, secure routing, and other network activities. Logical and computational trust analysis and evaluation are deployed among network nodes. Each node's evaluation of trust on other nodes is based on serious study and inference from trust factors such as experience statistics, data value, intrusion detection results, and references to other nodes, as well as a node owner's preference and policy. Ren et al. describe a technique to establish sufficient trust relationships in ad hoc networks with minimum local storage capacity requirements on the mobile nodes [185]. The authors propose a probabilistic solution based on a distributed trust model. A secret dealer is introduced only in the system bootstrapping phase to complement the assumption in trust initialization. With the help of the secret dealer, much shorter and more robust trust chains are able to be constructed with high probability. A fully self-organized trust establishment approach is then adopted to conform to the dynamic membership changes. But the shortcoming of this approach for the common sensor network is that it is not reasonable to introduce a dealer in a totally decentralized ad hoc environment.

The approaches described above are proposed in the context of ad hoc network. For the wireless sensor network, they can not be employed directly because of the capacity of the sensor. Some researchers specifically focus on the sensor networks that have been proposed recently. Ganeriwal and Srivastava propose a reputation-based framework for high integrity sensor networks [93]. Within this framework the authors employ a beta reputation system for reputation representation, updates, and integration. Tanachaiwiwat et al. [213] propose a mechanism of location-centric isolation of misbehavior and trust routing in sensor networks. In their trust model, the trustworthiness value is derived from the capacity of the cryptography, availability and packet forwarding. If the trust value is below a specific trust threshold, then this location is considered insecure and is avoided when forwarding packets. Liang and Shi focus on trust model developing and the analysis of rating aggregation algorithms in the open untrusted environment [140, 141]. Their findings and observations can be applied to wireless sensor networks directly, although the work is performed in the context of peer-to-peer settings. They propose a personalized trust model called PET in [141], which supports the customization of trustworthiness from the view of individual sensors. They find that the rating is not always helpful given the limitations of other factors. In the open environment with high dynamics the rating performance degrades and can produce negative effects. They observe that the storage space for saving self-knowledge is a potential bottleneck to the effect of ratings. Their recent simulation results show that it is better to treat the ratings from different evaluators equally given the dynamics of the open environment, and simply averaging ratings is appropriate considering the simplicity of the algorithm design and the low cost in running the system. They argue that the most important issue for building a trust model is adjusting parameters according to environment changes. These suggestions are quite useful for building trust models in the wireless sensor network given their simplicity and cost savings.

1.3 Book Contributions

The contribution of this book can be summarized in the following main points.

Contribution 1: Pair-wise key. The first research problem addressed is the node pair-wise authentication and the node pair-wise communication confidentiality. These are some of the fundamental challenges for this type of network. In fact, the constrained resources of a node make it difficult to solve this problem: Well known solutions adopted for wired networks cannot be used in WSN environment. Furthermore, as also shown later in this book, confidentiality and authentication are basic building blocks to face different security threats. We explored the various key distribution schemes proposed in the literature. They can be mainly classified as deterministic or probabilistic. Once we understood the limitations of the deterministic approaches we concentrated our efforts on the probabilistic schemes. Finally, we designed a new probabilistic solution, the Enhanced Cooperative Channel Establishment (ECCE) Protocol. We compared the performance of ECCE with the most known concurrent schemes via both analysis and simulations. The results showed that the ECCE Protocol presents higher probability for any pair of nodes to establish a secure channel and a higher resilience rate (i.e. the attacker needs a bigger effort to corrupt the channel). A preliminary version of the ECCE Protocol has been published in the *Second IEEE International Workshop on Sensor Networks and Systems for Pervasive Computing (PerSeNS 2006)* [46]. The final version of the protocol has been published in the journal *Ad Hoc Networks (Elsevier)* [47]. This contribution is described in Chap. 2.

Contribution 2: Capture Detection Protocols. Once studied the problem of authentication and communication confidentiality between nodes on the base of node's pre-deployed secret material, the next step has been to face the following scenario. The attacker physically captures a sensor node and tampers with it to read the keys it stores in its memory. In this way the attacker is able to know some keys that can be used to secure the communications in the network. This implies a threat to the authentication and the communication confidentiality. Furthermore, once captured a node, an attacker can perform other types of attacks. An example is the clone attack. That is, the attacker physically captures a network node and makes several clones of the honest node. Eventually, the attacker can maliciously reprogram the cloned nodes while keeping the pre-deployed secret material that the nodes can use to prove that their identities are "honest". So, the attacker can use the cloned nodes to perform malicious activities.

The node capture is the first step for an attacker to perform several other attacks that are crucial for WSNs. If we were able to detect the physical capture of a node we would be also able to avoid any subsequent attack such as the clone attack or the confidentiality violation. So, being able to deal with the capture detection means early detection and prevention of many important attacks.

Given the importance of this problem we focused our attention over the detection of the physical capture of a node. The literature did non present any work in the context of sensor networks. Our approach has been to leverage the network mobility in order for the nodes to trace the presence of the other nodes. The seminal idea has been published in the *First ACM Conference on Wireless Network Security (WiSec 2008)* [53]. In this framework we further proposed two protocols. We analysed the protocols and performed an extensive set of simulations, comparing the proposed protocols with a naïve solution that does not leverage the network mobility (i.e. it just exploits the classic message exchange). The results showed that the newly proposed solutions can be practically implemented in sensor networks and under certain mobility conditions (e.g. a certain average node speed) they perform better than solutions that do not leverage the network mobility. The resulting work can be found in [49]. This contribution is described in Chap. 3.

Contribution 3: Clone Detection Protocols. The proposed protocols for the capture attack detection present an increasing energy overhead while we require a performance improvement (i.e. early detection). If a node capture is not detected there can be several outcomes. In particular, we concentrated on the node cloning attack that seemed to be a new investigation area. In fact, when we started working on this problem there was just one theoretical solution [169]. We started studying this solution that actually resulted to be non practical for WSNs, as we investigated in our first work on this topic [52]. Following the stated properties that a distributed clone detection protocol should possess [52] we designed a Randomized, Efficient and Distributed (RED) protocol for the detection of the node replication attack. The work has been published at the *Eighth ACM International Symposium on Mobile Ad Hoc Networking and Computing (MobiHoc 2007)* [48]. A detailed analysis of the RED Protocol can be found in [55]. Finally, we also proposed a clone detection protocol that is particularly suitable for networks where a key predistribution mechanism is already implemented. This protocol has been published in the journal *Information Fusion (Elsevier)* [54]. This contribution is described in Chap. 4.

Contribution 4: Secure Aggregation. All the protocols previously discussed present a trade-off between effectiveness and efficiency. As an example, an early detection of a capture detection or a clone detection attack means a higher cost of the protocol in terms of energy consumptions. Because of the constrained resources of a WSN, the protocol energy consumption becomes prohibitive when it is required to reach given performances. That is, after this threshold (corresponding to some protocol's parameters setting) the attacker has some possibilities to succeed. This makes worth investigating the consequences of the attacker presence from a different point of view. The question becomes: Can a WSN service be resilient to a possible presence of an attacker? We tried to answer this question in the context of a typical service of WSN: Data aggregation. Assume a WSN is deployed to sense some environmental data (temperature, sound, etc.). Due to the constrained resources of WSNs we cannot imagine that each node sends its own sensed data to a collecting point, e.g. the Base Station (BS). To avoid this waste of energy the data aggregation comes in our help: Data are aggregated along the path to the BS, accordingly to the query. As an example, imagine the BS asks for the maximum value of temperature sensed within the network. Using a data aggregation protocol only the maximum value encountered during the aggregation will be forwarded.

The security problems of data aggregation mechanisms have been studied during the period the candidate spent as a Visiting Researcher at the Center for Secure Information Systems (CSIS) at George Mason University. In particular, the first problem addressed has been the secure computation of the Median aggregate. The result of this work has been published at the *Fourh International Conference on Security and Privacy in Communication Networks (SecureComm 2008)* [190]. This contribution is described in Chap. 4.

Contribution 5: Privacy in Data Aggregation. Another contribution on data aggregation in sensor networks, also given during the period at CSIS at George Mason University, is related to the privacy of a single node during the data aggregation. In many sensor network applications the data sensed by a single node can be related to a user (or a number of users): Information on patients health in hospital, water consumption in a city, etc. Then, in order to protect the people's privacy the data aggregation protocol that works in this type of context must protect the privacy of each single node. In particular, it should not be possible to relate a given sensed data to a given sensor node. We presented the first data aggregation protocol that guarantees the privacy of a node not only against the other nodes but also against the Base Station, which is the entity that eventually collects the aggregated data. Results are summarized in [60]. This contribution is described in Chap. 5.

1.4 Book Overview

In this book we discuss five fundamental mechanisms to build secure WSN. In particular, we start from the security issues related to a single node, that is (i) we deal with the authentication and communication confidentiality with other nodes. Then, we focus on network security, providing solutions for (ii) the node capture attack and (iii) the clone attack. Finally, we address security for a common WSN service: The data aggregation, providing solutions that (iv) are resilient to the attacker presence; and, (v) protect the privacy of the nodes.

The rest of this book is organized as follows.

Chapter 2 presents the ECCE Protocol: A new cooperative pair-wise key-establishment protocol for WSNs. Analysis and extensive simulations show that the proposed protocol presents an higher probability for two nodes to build a secure channel, i.e. the capability for these nodes to exchange data in a confidential way. Furthermore, the proposed protocol is shown to be more resilient to attacks: An attacker has to capture an higher number of nodes, when compared to competing protocols, in order to be able to threat the security of a given channel.

Chapter 3 introduces a new framework for the detection of a node physical capture in a WSN. In particular, we present a new approach for the detection of the capture attack considering a mobile WSN. The new approach is completely distributed and it is based on nodes mobility and nodes cooperation. The study of this approach shows that it is feasible for WSNs. Furthermore, we show that the proposed approach performs better compared to a classical approach, that we consider as a benchmark, based just on multi-hop message exchange between network nodes.

Chapter 4 studies a further security level assuming the attacker managed to physically capture a network node without being detected. In this chapter we focus on the detection of a clone attack: i.e. the attacker, after capturing a node, makes clones out of the honest node and uses the clones, re-inserted in the network, to perform malicious activities.

Chapter 5 investigates the security of data aggregation—a typical WSN service. In particular, we present a secure protocol for the computation of the Median of all of the nodes' sensed values. To the best of our knowledge this is the first secure protocol for the Median computation in WSNs.

Chapter 6 proposes a privacy-preserving aggregation protocol. Here we assume the attacker's aim is to violate the privacy of the data sensed by a single node (i.e. the attacker wants to correlate a data to a given node). To the best of our knowledge this is the first privacy-preserving data aggregation protocol that guarantees the node privacy not only against other nodes but also against the Base Station that wants to collect the aggregated data.

Chapter 7 concludes the book and outlines some directions that can be followed to continue researching the exposed problems.

Chapter 2
Pair-Wise Key Establishment

In this chapter we start addressing the security of Wireless Sensor Networks (WSNs) from the point of view which is closest to the single node: We consider the authentication and the confidentiality of the communications with other nodes. In particular, this chapter presents the ECCE Protocol, a new distributed, probabilistic, cooperative protocol to establish a secure pair-wise communication channel between any pair of sensors in a WSN. The main contribution of the ECCE Protocol is: To allow the set up of a secure channel between two sensors (principals) that do not share any pre-deployed key. This feature is obtained involving a set of sensors (cooperators) in the channel establishment protocol to provide probabilistic authentication of the principals as well as the cooperators. In particular, the probability for the attacker to break authentication check decreases exponentially with the number of cooperators involved. We provide analytical analysis and extensive simulations of the ECCE, which show that the proposed solution increases both the probability of a secure channel set up and the probability of channel resilience with respect to other protocols.

2.1 Introduction

WSNs are expected to be the basic building block of pervasive computing environments, hence establishing secure pair-wise communications could be useful for many applications. In particular, it is a pre-requisite for the implementation of secure routing, and can be useful for secure group communications. Further, pair-wise secure communication allow in-network processing [247], or facilitate the establishment of a cluster key, hence enabling *passive participation*, in which a sensor node can take certain actions based on overheard messages. It was pointed out in [3, 36, 173], that asymmetric cryptography such as RSA or Elliptic Curve Cryptography (ECC) is unsuitable for most sensor architectures due to high energy consumption and

© Springer Science+Business Media New York 2016
M. Conti, *Secure Wireless Sensor Networks*, Advances in Information Security 65,
DOI 10.1007/978-1-4939-3460-7_2

increased code storage requirements. However, it is worth noticing that evolution in technology allows to sparingly use asymmetric cryptography for a certain class of WSN [222]. For instance, in [169] the authors devise a protocol that, with the seldom use of ECC, thwarts the replication attack [79, 159]. However, it seems reasonable that there will be always some classes of WSNs in which asymmetric cryptography would rise an unfeasible cost due to either energy consumption or memory constraints. Indeed, as for energy consumption, in a mobile WSN if we have a secure key establishment protocol based on ECC whenever two sensors want to agree on a shared key for the first time, this would put high requirements on battery consumption. As for memory, it seems unfeasible that any node could host the public keys of all the other nodes in the network (for instance, the Mica mote is equipped with a 8 bit 4 MHz processor and has 4 KB of RAM and 128 KB of flash RAM only). Note that the constraint on memory stands even in a static WSN. Hence, while solutions that intend to address specific problems can directly benefit of the sparingly use of ECC [169], building communication channel based on symmetric algorithms, which are three order of magnitude more efficient then ECC [222], is still an attractive research field [36, 197].

This chapter presents the ECCE Protocol, a new protocol to establish a secure pair-wise communication channel between any pair of sensors in the WSN. The ECCE Protocol can be classified as probabilistic and cooperative. Unlike other protocols for channel establishment, ECCE allows to establish a secure channel between sensors that do not share any key, involving a set of cooperating sensors (cooperators) which are not required to share a key with both principals. The same feature is not present in actual protocols such as Multipath Key Reinforcement [37] and Cooperative [75]. The overhead required is limited and it is sustained just once during the sensor life-time. ECCE shows better performance in channel existence and channel resilience than existing protocols. The Protocol also guarantees implicit and probabilistic mutual authentication of principals and cooperators without any additional overhead and without the presence of a base station. Further, the proposed protocol could be used also between sensor that already share some secret keys to increase the resilience of these shared keys. The proposed protocol is also adaptive to the required security level: To achieve an higher level of security, it suffices to involve an higher number of cooperators in the channel set-up. Finally, the protocol allows to trade off the memory required to store pre-deployed keys with cooperators. In particular, it is possible to set the number of cooperators in order to have a reduced key ring that provides the same level of security and the same probability of channel existence of solutions that involve no cooperators but a large key ring size. For example, choosing a pool of size 1,000, a key ring of size 12, and involving 8 cooperators, provides the same probability of channel existence of a scenario in which every sensor has 20 pre-deployed keys but there are no cooperators. As for resiliency, with a pool of size 10,000, a key ring of size 100 and 8 cooperating sensors, the attacker is required to capture 110 sensors to corrupt a channel, while with the same parameters, but with no cooperators, the attacker has to corrupt only 75 sensors to corrupt the channel. Note that reducing the key ring size provides the possibility for sensors to store the cooperative keys set-up with the ECCE Protocol. Compared to several recently proposed approaches such as [11, 66, 196] which fall in different categories of key establishment schemes, our analytical and experimental results show that the ECCE Protocol has better performance than the other protocols as for channel existence and channel resiliency to the attacker.

Organization

The remainder of this chapter is organized as follows. In Sect. 2.2, we review the current contributions in the field. In Sect. 2.3, we report some preliminaries and define our system assumptions. In Sect. 2.4, we describe the ECCE Protocol, while in Sect. 2.5, we analyze the probability to establish a secure channel and the resilience of the established channel.

2.2 Related Work

Some research focus on key establishment protocol for WSN based on centralized solution. Examples of centralized protocols include [73, 143, 173]. Centralized protocols assume the presence of a Base Station (BS), which takes part in the process of establishing a pair-wise key between pairs of sensors. This kind of solution has some drawbacks, for instance the energy consumption experienced by the nodes close to the BS, and the presence of a single point of failure. Other research focus on distributed solution for pair-wise keys establishment. To better refine this classification, we can distinguish between *deterministic* and *probabilistic* solutions. As for deterministic solutions one can see [36, 144, 197]. However, each of these solutions suffer of a specific type of problem. In [144], the attacker only needs to corrupt a constant number of nodes to disrupt the confidentiality of the whole network. In [197], the authors recognise that given a fixed key-ring size, this limits the number of sensors in the network. Finally, in [36], each sensor is required to store $O(\sqrt{N})$ keys; moreover, the number of sensors that belong to the same WSN (that is N) must be known at design time. As for probabilistic solutions, the idea of probabilistic key sharing for WSN was firstly introduced in [85]. In the proposed solution, each of the N sensors of the WSN is assigned K symmetric encryption keys randomly selected without replacement from a common Pool of P keys (*key pre-deployment phase*). When two sensors need to communicate securely, they must first find out which keys (if any) of the Pool they share (*shared-key discovery phase*). Then, they compute a common key as a function of the shared keys (*pairwise-key establishment phase*). This latter key is used to secure the channel by using a symmetric key encryption algorithm. Some solutions based on pseudo-random key assignment are presented in [37, 74, 75, 80, 144, 249]. However, these solutions show limited resiliency to tampering, as highlighted in [74, 75], or just require cooperating nodes to share a key with both principals. For the shared-key discovery phase different mechanisms have been proposed. In [85], the challenge-response and key index notification are proposed. With K keys stored in each sensor, the challenge-response requires sending and receiving K messages, to perform K encryption and, in the worst case, K^2 decryption. With the pseudo-random key index transformation proposed in [249], no message exchange are needed between sensors that want to establish a secure channel. Moreover, the attacker can compute the IDs of the keys stored by each sensor just acquiring the sensor ID; this solution shows a weakness similar to that in [85] and above exposed. The problem related to the information leakage of the keys' ID is solved in [74, 75]. Here the authors introduce a mechanism (ESP) that requires no message exchange for the shared-key discovery phase and reveals to the attacker no information about the keys it does not hold yet. Further, ESP provides probabilistic node authentication: A sensor can prove its identity by proving knowledge of the keys it is supposed to hold. The *Efficient and Secure Pre-deployment (ESP) scheme* works as follows. Consider a sensor a. For every key k_i^P of the pool, compute $z = f_y(a \parallel k_i^P)$, where f_y is a *pseudo-random function*, that is an efficient (deterministic) algorithm which given an h-bit seed, y, and an h-bit argument, x, returns an h-bit string, denoted $f_y(x)$, so that it is infeasible to distinguish the responses of f_y, for a uniformly chosen y, from the responses of a truly random function. Then, put k_i^P into the key ring of a, if and only if $z \equiv 0 \mod (|P|/K)$. ESP supports a very efficient key discovery procedure. Consider a sensor b that is willing to know which keys it shares with sensor a. For every key k_j^b in the key ring of b sensor b computes $z = f_y(a \parallel k_j^b)$. Then, by testing $z \equiv 0 \mod (|P|/K)$, b discovers whether sensor a also has key k_j^b or not. Indeed, whoever already knows key k_i^P is the only one who can know whether k_i^P is in the key ring of a or not. This is computationally impossible for all other entities, since f_y, being a pseudo-random function, is also one-way and thus hard to invert [99]. For this reason, from the ID of a node an attacker cannot acquire neither the keys stored by this node, nor the corresponding key indexes: $f_y(x)$ is applied to the actual value of the key, not to the corresponding key index. This kind of ID-based security could be thwarted only with the random capture of a large number of nodes (as shown later in this chapter) or via a node replication attack [52]. Finally, it is important to note that another property a WSN is required to enforce is connectivity. The connectivity problem was initially addressed in [85]; however, Ganeriwal et al. [76] revised and extended the model of connectivity in WSN.

2.3 Preliminaries and Assumptions

This section reports the notation and the assumptions that will be used in the following. For clarity, in Table 2.1 we list the symbols used in the chapter.

2.3.1 Security Requirements and Threat Model

We assume the following working hypothesis:

- Communication infrastructure: We assume an underlying routing mechanism such that any node can send a message (leveraging multi-hop) to any other node in the

Table 2.1 Pair-wise key establishment: Notations

Symbol	Meaning
N	Number of sensors in the WSN
P	Size of the pool from which the keys are drawn
K	Number of keys assigned to each sensor (key-ring size)
$\mathscr{C}$	Set of cooperating sensors
a, b	Principals
c_i	ith cooperating sensor, where $1 \leq i \leq |\mathscr{C}|$
w	Number of corrupted sensor in the WSN
k_i^a	ith key assigned to sensor a, where $1 < i < k$
$k_{h,l}$	Key between sensors h and l
$K_{h,l}$	Key computed with Direct Protocol [75], if possible, string of 0_s otherwise
$K_{h,l}^{\mathscr{C}}$	Key computed with Cooperative protocol [75]
$\bar{K}_{h,l}^{\mathscr{C}}$	Key computed with ECCE Protocol
$E_k(x)$	Encryption of string x with key k
$D_k(x) = E_k^{-1}(x)$	Decryption of string x with key k
$H(x)$	Hash function
$DH(x)$	Hash function doublehash, $DH(x) : \{0, 1\}^{|x|} \rightarrow \{0, 1\}^{2|x|}$
$LS(x)$	Less significant bits of string x, where $|LS(x)| = \frac{|x|}{2}$
$MS(x)$	Most significant bits of string x, where $|MS(x)| = \frac{|x|}{2}$

network. An example of such a globally addressable communications infrastructure is in [31];

- Sensors are randomly scattered in an unattended and often adversarial environments. To preserve their low cost, as well as to save power [9], they are not tamper proof [247]. Hence, we can assume that sensors can be physically captured and the attacker can acquire all the information stored within captured sensors;
- Good security engineering practice [8]: The algorithms, protocols and mechanisms that are employed to secure the WSN are publicly known. Only the keys in the sensors' key rings and in the Pool are initially secret. Moreover, the cryptographic primitives that are employed are at least computationally secure;
- Attacker model: We assume the strong node-compromise attacker model adopted in [36]. Specifically, we assume that the attacker is capable of compromising a fraction of the total number of nodes in the network and exposing the secret information contained within them. There can be two forms of node compromise.

In *passive node compromise* we assume that after node compromise, the attacker can only launch passive attacks such as eavesdropping. In *active node compromise* we assume that the attacker is also able to perform active attacks such as providing false routing metrics through the compromised node. We assume that the goal of the adversary is the exposure of the keys stored by the sensors, including those established with the ECCE Protocol.

- Channel establishment: The ECCE Protocol requires three phases, that is, key pre-distribution, shared-key discovery and channel establishment. We assume keys are assigned to sensors according to the ESP procedure [74], hence the first two phases are carried out as described in Sect. 2.2.

  As for channel establishment, we will cope with this issue exploiting cooperating sensors, as described in Sect. 2.4. In the remainder of this chapter we assume that two sensors sharing one or more pre-deployed keys can compute a shared key via the Direct Protocol [75], that is the channel is built combining the keys the two principals share. The existence of a key established via the Direct Protocol translates into the existence of a Direct channel (that is a *link*) between the two sensors. Finally, we will use the term *corrupted* channel to refer either the fact that the keys the channel is built with are known to the attacker, or that the channel does not exist (that is the sensors do not share any key).

2.4 The ECCE Protocol

The ECCE Protocol involves, beyond the principals, other sensors (cooperating sensors). It is based on the fact that each distinguished cooperating sensor c_i can efficiently compute the keys it shares with each other cooperator. This can be efficiently done assuming that the key pre-deployment procedure is carried out according to ESP scheme, detailed in Sect. 2.2. Based on the keys cooperating sensor c_i shares with each other cooperator, c_i can compute $(s_1, \ldots, s_{i-1}, s_{i+1}, \ldots, s_{|\mathscr{C}|-1})$. These shared information are further combined to compute two values (v_1, v_2). Each value is then sent to the two principals a and b. Each principal, upon receiving all the values provided by each cooperator, computes the key $\bar{K}_{a,b}^{\mathscr{C}}$ that will be employed to secure communication between the two principals. The details of the protocol follow.

If sensor a wants to establish a secure channel with sensor b, a chooses a set $\mathscr{C} = \{c_1, \ldots, c_m\}$ of cooperating sensors such that $a, b \notin \mathscr{C}$ and $m \geq 1$. Then, a sends a request of cooperation to c_i, for each $c_i \in \mathscr{C}$. If a Direct key K_{a,c_i} between a and c_i exists, the request of cooperation is sent encrypted with K_{a,c_i}, else the request is sent not encrypted. The request carries the ID of b and the IDs of the sensors in $\mathscr{C}$.

Each sensor computes two different values (v_1, v_2), v_1 to be sent to the sensor a and v_2 to be sent to the sensor b. In particular, every cooperating sensor c_i computes the value v_1 to be sent to sensor a as follows. The key $K_{c_i,b}$ is computed and then hashed with the ID of a (ID_a): $H(ID_a, K_{c_i,b})$. The keys shared with all other cooperating sensors are computed via the ESP protocol. For each cooperating sensor c_j where $ID_{c_i} < ID_{c_j}$, the hash $H(ID_a \oplus ID_b, LS(DH(K_{c_i,c_j})))$ is computed; for each cooperating sensor c_j, where $ID_{c_i} > ID_{c_j}$, the hash $H(ID_a \oplus ID_b, MS(DH(K_{c_i,c_j})))$ is computed. The XOR of all the computed hash is executed and the resulting string v_1 is sent to a encrypted with Direct key $K_{c_i,a}$. Note that, as exposed in Table 2.1, we have assumed that the Direct key procedure always returns a key. This key is $K_{c_i,a}$ if it exists, or an appropriate string of $0s$ -a publicly known key- otherwise.

Every cooperating sensor c_i computes the value v_2 to be sent to sensor b as follows. Key $K_{c_i,a}$ is computed and then hashed with the ID of b: $H(ID_b, K_{c_i,a})$. The keys shared with all other cooperating sensors are computed and for each cooperating sensor c_j satisfying $ID_{c_i} < ID_{c_j}$ the hash $H(ID_a \oplus ID_b, MS(DH(K_{c_i,c_j})))$ is computed. For all cooperating sensors c_j, with $ID_{c_i} > ID_{c_j}$, the hash $H(ID_a \oplus ID_b, LS(DH(K_{c_i,c_j})))$ is computed. The XOR of all the computed hash are executed and the resulting value v_2 is sent to b encrypted with Direct key $K_{c_i,b}$.

For every cooperating sensor c_i from which sensor a receives a reply message v_1 before time-out expires (let $\mathscr{C}_r$ the set of replying cooperating sensors), sensor a computes the hash $H(ID_b, K_{a,c_i})$ and then $g_i = v_1 \oplus H(ID_b, K_{a,c_i})$. When a either receives all the reply messages from the cooperating sensors, or the last time-out expires, a computes the ECCE key as follows: $\bar{K}^{\mathscr{C}}_{a,b} = K_{a,b} \bigoplus_{i=1}^{|\mathscr{C}_r|} g_i$. The ECCE key $\bar{K}^{\mathscr{C}}_{a,b}$ is finally hashed and sent to b. The hashed ECCE key, when received by b could be used by b to check whether the locally computed ECCE key matches the ECCE key computed by a. Sensor b can set a time-out to limit the delay of expected messages. When the time out expires, b computes the ECCE key in a way similar to a, and will finally check whether the hash of the ECCE key received by a matches with the hash of the ECCE key locally computed.

Algorithm 1 shows the detailed pseudo-code of the ECCE Protocol. Algorithms 2 and 3 illustrate the functions COMPUTEMSGFORSOURCE and COMPUTEMSG-FORDESTINATION used by every cooperating sensor to compute the message to be sent to a and b respectively. The H function used in the Protocol is employed to produce a non-invertible image of the keys, to avoid information leakage [37, 75].The algorithm SELECTRANDOM selects the cooperating sensors in a pseudo random fashion among all the possible sensors in the network. However, note that this choice of cooperators could be unsatisfactory. For instance, the number of cooperators that share a key with the principals or with other cooperators might be small. This could affect the channel resilience, as we will see in Sect. 2.5.2. To cope with this problem, we could select cooperators according to some other policy. For instance, we could accept a selected cooperator only if it shares a Direct key with both principals. Note that it is not required that the key shared with principal a is the same key shared with principal b.

Involving cooperators allow the ECCE Protocol to be adaptive to different security requirements: If it is required to increase the channel resilience, than this objective can be achieved involving more cooperating sensors. The use of cooperators in the ECCE Protocol further allows to balance the burden of protocol execution among all the cooperators and the principals. Indeed, each cooperator computes $|\mathscr{C}| + 1$ Direct keys and hash, while sending only 2 messages. If principal a wants to set up an ECCE key with principal b, a has to forward $|\mathscr{C}| + 1$ messages and to receive $|\mathscr{C}|$ messages. Sensor b only needs to collect the $|\mathscr{C}|$ messages sent to it by the cooperating sensors.

If a cooperating sensor in $\mathscr{C}$ is not available (for instance, due to a node failure), using the time-out the protocol will not fail or deadlock, granting sensor failure resilience. Observe also that if there is not a Direct key between some of the cooperators involved in the protocol, this does not imply the failure of the protocol.

Input   : b : ID of the receiving sensor.
Output: $\bar{K}_{a,b}^{\mathscr{C}}$

1 **begin**
2      $\mathscr{C} = \text{SELECTRANDOM}(NeededCooperators)$;
3   *Set time-out* Δ ;
4      $\bar{K}_{a,b}^{\mathscr{C}} = 0_s$;
5   **forall the** $c_i \in \mathscr{C}$ **do**
6          $K_{a,c_i} = \text{DIRECT\_ PROTOCOL}(c_i)$;
7          $a \rightarrow c_i :< a, c_i, E_{K_{a,c_i}} (req\_coop \parallel C \parallel b) >$;
8   **end**
9      $\mathscr{C}' = \mathscr{C}$;
10   **while** $\mathscr{C}' \neq \emptyset$ **and** (**not** *elapsed*(Δ)) **do**
11          $a \leftarrow c_i :< c_i, a, E_{K_{a,c_i}} (b, \text{COMPUTEMSGFORSOURCE}(a, b, c_i, C)) >$;
12          $b \leftarrow c_i :< c_i, b, E_{K_{c_i,b}} (a, \text{COMPUTEMSGFORDESTINATION}(a, b, c_i, C)) >$;
13          $s = E_{K_{a,c_i}}^{-1} \left(E_{K_{a,c_i}} (\text{COMPUTEMSGFORSOURCE}(a, b, c_i, C')) \right)$;
14          $\mathscr{C}' = \mathscr{C}' - \{c_i\}$;
15          $\bar{K}_{a,b}^{\mathscr{C}} = \bar{K}_{a,b}^{\mathscr{C}} \oplus s \oplus H \left(ID_b, K_{a,c_j} \right)$;
16   **end**
17      $\bar{K}_{a,b}^{\mathscr{C}} = K_{a,b} \oplus \bar{K}_{a,b}^{\mathscr{C}}$;
18      $a \rightarrow b :< a, b, H \left(E_{\bar{K}_{a,b}^{\mathscr{C}}} \right) >$;
19 **end**

Algorithm 1: ECCE Protocol

Input   : a : ID of the sensor (cooperator). b : ID of the source. c_i : ID of the receiving
        sensor. $\mathscr{C}$: Set of IDs of the others cooperating sensors.
Output: Compute the message that the cooperator send to source sensor

1 **begin**
2      $Msg = H \left(ID_a, K_{c_i,b} \right)$;
3   **forall the** $c_j \in C_{i'} (i \neq j)$ **do**
4   **if** $\left(ID_{c_i} < ID_{c_j} \right)$ **then**
5             $Msg = Msg \oplus H \left(ID_a \oplus ID_b, LS \left(DH \left(K_{c_i,c_j} \right) \right) \right)$;
6   **end**
7   **if** $\left(ID_{c_i} > ID_{c_j} \right)$ **then**
8             $Msg = Msg \oplus H \left(ID_a \oplus ID_b, MS \left(DH \left(K_{c_i,c_j} \right) \right) \right)$;
9   **end**
10   **end**
11 **end**

Algorithm 2: COMPUTEMSGFORSOURCE

Input : a : ID of the sensor (cooperator). b : ID of the source. c_i : ID of the receiving
        sensor. $\mathscr{C}$: Set of IDs of the others cooperating sensors.
Output: Compute the message that the cooperator send to receiving sensor

1 **begin**
2 | $Msg = H\left(ID_b, K_{c_i,a}\right)$;
3 | **forall the** $c_j \in C_{i'}(i \neq j)$ **do**
4 | | **if** $\left(ID_{c_i} < ID_{c_j}\right)$ **then**
5 | | | $Msg = Msg \oplus H\left(ID_a \oplus ID_b, MS\left(DH\left(K_{c_i,c_j}\right)\right)\right)$;
6 | | **end**
7 | | **if** $\left(ID_{c_i} > ID_{c_j}\right)$ **then**
8 | | | $Msg = Msg \oplus H\left(ID_a \oplus ID_b, LS\left(DH\left(K_{c_i,c_j}\right)\right)\right)$;
9 | | **end**
10 | **end**
11 **end**

Algorithm 3: COMPUTEMSGFORDESTINATION

Sending the list of all cooperators to each $c_i \in \mathscr{C}$ can help the attacker: It will be sufficient to corrupt a channel between the sender and one of the cooperators, and the set of cooperators would be disclosed. However, employing the ESP mechanism, the attacker cannot know the set of keys the cooperators hold, other than the subset of keys it already knows. Hence, the attacker is still forced to corrupt cooperating nodes if it wants to reduce its efforts [75]. Further, to decrease the possibility for the attacker to acquire the list of all the cooperators, the set $\mathscr{C}$ of cooperators could be partitioned in subset $\mathscr{C}_1, \ldots, \mathscr{C}_q$: Corrupting a channel or a cooperator within a specific subset $\mathscr{C}_i$ does not reveal any information about the cooperators belonging to other subsets. This countermeasure has been implemented in a version of the ECCE Protocol that we will refer to as Partitioned ECCE.

2.5 Security Analysis

The condition that must be verified to guarantee the confidentiality of keys set-up using the ECCE Protocol, is the existence of a non corrupted path between the principals a and b $(a - b)$, where each link of this path is built with a Direct key and the intermediate nodes between a and b are the cooperating sensors. As an example, in Fig. 2.1 the sensors a and b use the ECCE Protocol to build a confidential key. In Fig. 2.1 the path composed of continuous lines signals a Direct key unknown to the attacker, while the dashed line signals a corrupted link or a non existing channel. In this example, the confidentiality of the established key $\bar{K}_{a,b}^{\mathscr{C}}$ is guaranteed by the existence of the path (a, c_1, c_2, c_3, b). If the attacker does not hold the Direct keys used to secure the links (a, c_1), (c_1, c_2), (c_2, c_3), and (c_3, b) there is no way to build the ECCE key.

Fig. 2.1 Example of ECCE channel not corrupted: Existence of path (a, c_1, c_2, c_3, b)

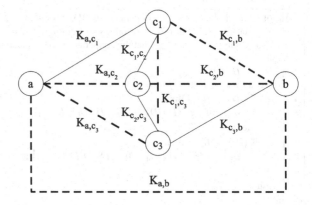

2.5.1 Channel Existence

In this section we analyze the probability that a pair of sensors succeeds to establish a confidential key using the ECCE Protocol. This probability depends on the probability of existence of Direct keys shared between all the possible pairs of sensors in $\mathscr{C} \cup \{a\} \cup \{b\}$. The probability to establish a Direct key is given by the probability that two sensors share at least one of the assigned keys of the Pool. From [85], it follows that:

$$\Pr[\text{link exists}] = 1 - \frac{\binom{P-K}{K}}{\binom{P}{K}} = 1 - \frac{K!(P-K)!(P-K)!}{P!K!(P-2K)!} \qquad (2.1)$$

In the following to ease exposition, we assume that the existence of each link is independent from each other [37, 85]. We indicate with p the probability of existence of a single link. Further, $path_{ECCE}(|\mathscr{C}|)$ represents the event of a path between principals (that can be established also via a direct link between principals), while $path_{ECCE}(|\mathscr{C}|, nodir)$ accounts for the event of a path between principals but note that this path can be formed only through cooperating sensors (it is assumed that the direct link between principals does not exist). Then, we have:

$$\Pr[path_{ECCE}(|\mathscr{C}|)] = p + (1-p)\Pr[path_{ECCE}(|\mathscr{C}|, nodir)] \qquad (2.2)$$

The existence of a not corrupted direct link between the principals implies the existence of a non corrupted ECCE channel. Should this direct link do not exist, then the existence probability of a not corrupted channel is equal to the probability that at least one non corrupted path involving cooperating sensors do exists.

To compute this probability, take into consideration all the possible links of type (a, c_i) (grouped in the set l_{src}) and (c_i, b) (grouped in the set l_{dst}). Hence:

$$\Pr\left[path_{ECCE}(|\mathscr{C}|, nodir)\right]$$
$$= \sum_{A=0}^{|\mathscr{C}|} \sum_{B=0}^{|\mathscr{C}|} \Pr\left[|l_{src}| = A\right] \Pr\left[|l_{dst}| = B\right]$$
$$\cdot \left(\Pr\left[path_{ECCE}(|\mathscr{C}|, nodir) \mid |l_{src}| = A, |l_{dst}| = B\right]\right)$$
$$= \sum_{A=0}^{|\mathscr{C}|} \sum_{B=0}^{|\mathscr{C}|} p^A(1-p)^{|C|-A} p^B(1-p)^{|C|-B}$$
$$\cdot \left(\Pr\left[path_{ECCE}(|\mathscr{C}|, nodir) \mid |l_{src}| = A, |l_{dst}| = B\right]\right) \qquad (2.3)$$

We remark that when $A = 0$ we have a null probability of having a path between a and the set of cooperators; if $B = 0$ then there are no paths between b and the sensors in $\mathscr{C}$. Let $C(l_{src})$ and $C(l_{dst})$ be the sets of cooperating sensors in $\mathscr{C}$ that share a key with sensor a and b respectively. A path between the principals exists with probability 1 if $C(l_{src}) \cap C(l_{dst}) \neq \emptyset$. Fixing $|l_{src}| = A$ and $|l_{dst}| = B$ we have that:

$$\Pr[C(l_{src}) \cap C(l_{dst}) \neq \emptyset] = 1 - \frac{\binom{|\mathscr{C}|-A}{B}}{\binom{|\mathscr{C}|}{B}}$$

Equation. 2.3 can then be expressed as:

$$\sum_{A=0}^{|\mathscr{C}|} \sum_{B=0}^{|\mathscr{C}|} p^A(1-p)^{|C|-A} p^B(1-p)^{|C|-B} \cdot \left(\left(1 - \frac{\binom{|\mathscr{C}|-A}{B}}{\binom{|\mathscr{C}|}{B}}\right)\right.$$
$$\left. + \Pr[path_{ECCE}(|\mathscr{C}|, nodir) \mid C(l_{src}) \cap C(l_{dst}) = \emptyset, |l_{src}| = A, |l_{dst}| = B]\right)$$

$$(2.4)$$

We must now calculate the existence probability of the paths $(a - b)$ that use more than one cooperating sensor; see for instance Fig. 2.1. In order not to incur in problems of dependency when dealing with probability, we will consider a modified scheme of the ECCE Protocol. This scheme helps the attacker because it excludes some cases in which the path would exist, hence this analysis will provide a lower bound to the security provided. In particular, we will only use the paths that join cooperating sensors in $C(l_{src})$ (called s) to cooperating sensors in $C(l_{dst})$ (called d), of length 1 or 2. As an example, Figs. 2.2 and 2.3 show the paths $(s - d)$ of length

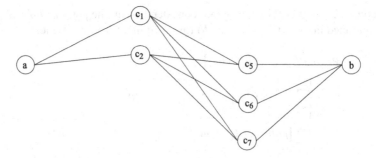

Fig. 2.2 Paths $(s - d)$ of length 1

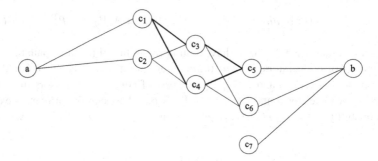

Fig. 2.3 Paths $(s - d)$ of length 2

1 and 2 respectively, for one particular choice of $\mathscr{C}$, l_{src} and l_{dst}. Finally, Eq. 2.2 can be expressed as:

$$\Pr\left[path_{ECCE}(|\mathscr{C}|)\right] = p + (1 - p)$$
$$\cdot \left(\sum_{A=0}^{|\mathscr{C}|}\sum_{B=0}^{|\mathscr{C}|} p^{A}(1-p)^{|C|-A}p^{B}(1-p)^{|C|-B} \cdot \left(\left(1 - \frac{\binom{|\mathscr{C}|-A}{B}}{\binom{|\mathscr{C}|}{B}}\right)\right.\right.$$
$$+ \left.\left.\left(1 - \left((1-p)^{AB}\left(1-p^2\right)^{(|C|-A-B)\cdot min\{A,B\}}\right)\right)\left(\frac{\binom{|\mathscr{C}|-A}{B}}{\binom{|\mathscr{C}|}{B}}\right)\right)\right)$$
$$(2.5)$$

These analytical results will be compared with simulation results in Sect. 2.6. The simulations performed support the behaviour predicted by the analytical model.

2.5.2 Channel Resilience

In this section we analyze the probability that the key established by the principals, using the ECCE Protocol, is not corrupted, that is not computable by the attacker through the information it holds. As discussed for the channel existence probability, also the resilience probability depends on the probability of resilience of the single Direct keys used in the construction of ECCE key. The probability that a Direct key is corrupted depends on the probability that a single key k_i of the Pool is corrupted. Considering w compromised sensors:

$$\Pr\left[key \; k_i \; is \; corrupted\right] = 1 - \left(1 - \frac{K}{P}\right)^w \tag{2.6}$$

If one knows the probability that a key is corrupt, then it could be possible to calculate the probability that an existing link is corrupted.

$$
\begin{aligned}
\Pr\left[link \; is \; corrupted \mid link \; exists\right] &= \frac{\Pr\left[link \; is \; corrupted \; \cap \; link \; exists\right]}{\Pr\left[link \; exists\right]} \\
&= \frac{\sum_{i=1}^{K} \left(\Pr\left[key \; is \; corrupted\right]\right)^i \Pr\left[i \; shared \; keys\right]}{\Pr\left[link \; exists\right]} \\
&= \frac{\sum_{i=1}^{K} \left(1 - \left(1 - \frac{K}{P}\right)^w\right)^i \frac{\binom{K}{i}\binom{P-K}{K-i}}{\binom{P}{K}}}{1 - \frac{\binom{P-K}{K}}{\binom{P}{K}}}
\end{aligned} \tag{2.7}
$$

Replacing the probability given by Eq. 2.7 in Eq. 2.5 (Eq. 2.7 gives the value for probability p), it is possible to obtain the probability that an ECCE channel is corrupted, assuming that all the pairs of cooperators share a Direct key. In a similar way, to assess the probability that an ECCE channel exists and is not corrupted, it is sufficient to replace the parameter p in Eq. 2.5 with the following formula:

$$\Pr\left[link \; exists\right] \cdot \Pr\left[link \; not \; corrupted \mid link \; exists\right]$$
$$= \Pr\left[link \; exists\right] \cdot \left(1 - \Pr\left[link \; corrupted \mid link \; exists\right]\right)$$

Again, in Sect. 2.6 we will show that simulation results support the derived analytical model.

2.5.3 Probabilistic Authentication

Another important issue of WSN security is node authentication. For instance, any scheme for the revocation of misbehaving nodes has its basis on the certainty of the nodes identities. Authentication can mitigate many dangerous attacks, like the replication of malicious sensors [79, 159]. In the following we discuss how the ECCE Protocol provides probabilistic authentication.

Based on the ECCE Protocol, every cooperating sensor c_i generates the messages to send by XORing the strings derived from the keys shared with the other cooperating sensors and the strings:

- $H(ID_a, K_{c_i,b})$, for the message destined to the sender sensor;
- $H(ID_b, K_{a,c_i})$, for the message destined to the receiving sensor.

Note that all cooperating sensors implicitly verify that both the principals have the keys that the principals should possess. This verification is possible due to the ESP mechanism. If a principal declares a false ID, cooperating sensors will use the keys that they should share with the sensor identified by ID. If just one of these keys is not possessed by the malicious principal that provided the fake ID the ECCE key cannot be established [74]. However, if the malicious principal possesses all the keys shared by the cooperators with the sensor identified by ID, the authentication process succeeds. For this reason the authentication is only probabilistic. Observe that, with respect to the Direct channel in which only one principal verifies the identity of the other party, in ECCE all the cooperating sensors verify the same identity with possibly different key-rings, hence the probability that a malicious sensor is not detected is smaller than in the Direct channel. In particular, since the authentication check performed by cooperators is carried out independently from each other, the probability that a fake principal succeeds in the authentication process, decreases exponentially with the number of cooperators involved. Further, note that the same mechanism supports authentication among the cooperating sensors as well. We remark that this authentication mechanism does not involve messages overhead other than the (limited) overhead required for the creation of the confidential channel.

2.6 Simulations and Discussion

In order to supply an experimental support to the analytical results developed in the previous section, we have performed extensive simulations. In particular, the ECCE Protocol has been compared with the following protocols:

- Direct [74];
- Cooperative [74];
- Extended Cooperative;
- MKR (Multipath Key Reinforcement) [37];
- Extended MKR;
- Partitioned ECCE (we have divided the set $\mathscr{C}$ in independent subsets of size 2, 3 and 4).

We assume that in all the considered protocols the ESP mechanism [74] is used in the *shared-key discovery phase*. We introduce the Extended version of both the

Cooperative and the MKR protocol, in which we assume that the existence of the Direct channel between the principals is not necessary. This optimization is based on the observation that, in the Cooperative and the MKR channel construction, the existence of the direct link between principals is only used to send some information to the receiving sensor; however, this information can be sent via a threshold scheme (t, c) [201] through the cooperating sensors.

We remark that a channel built according to the Cooperative Protocol, which requires that there is at least one shared key between sender and receiver, has the same existence probability of a channel established with the Direct Protocol. In fact, in both the Direct and the Cooperative Protocol, the necessary condition for the channel existence is the existence of the Direct link between the principals. Hence, the use of cooperating sensors in the Cooperative Protocol is only useful to increase confidentiality resiliency against the attacker, while the existence probability does not increase. This observation holds for the MKR Protocol as well.

The Extended Cooperative needs only a direct link or a 2-hop path between principals realized through the cooperators in $\mathscr{C}$ for the channel to exist. The behaviour of the Extended Multipath Key Reinforcement (MKR) [37], assuming a $2 - hop$ MKR scheme as in [37], provides the same probability of channel existence and channel resiliency as the Cooperative. In both ECCE and Extended MKR the necessary condition for channel existence is the existence of a direct link between the principals or a path through the cooperators in $\mathscr{C}$. Hence, we can state that the existence probability of ECCE and the Extended MKR is roughly the same but, as discussed in Sect. 2.5.2, this equivalence does not hold as for the resilience, where ECCE performs better.

Figure 2.4 compares the analytical and experimental results as for channel existence. We have fixed $P = 100, K = 5$, while $|\mathscr{C}|$ ranges from 0 to 12. As expected, the assumption of independence among the links, used to ease the analysis in Sect. 2.5, implies an upper bound on the estimation of channel existence, as simulation results show (*an. res.* refers to the analytical results of Sect. 2.5 while *sim.* refer to simulation results). However, the simplified model used to analytically study the behaviour of the ECCE Protocol did not take into consideration some cases in which a channel between cooperators could exist. For this reason, when more than 9 cooperating sensors are involved in the channel establishment, the analytical results for ECCE are superseded by the simulation results. However, note that the difference between the simulation and the analytical results between the two slopes is tiny for the whole range of cooperating sensors considered. Observe that using no cooperating sensors, the behaviour of the ECCE, the Cooperative and MKR Protocol is similar to that of the Direct Protocol, while increasing the number of cooperators, the existence probability of the cooperative protocols (Cooperative, MKR and ECCE) increases as well. In particular, the ECCE Protocol provides better channel existence probability, and this probability improves with the number of cooperators.

Figures 2.5, 2.6, and 2.7 plot the existence probability of a secure channel established with the ECCE protocol, together with the same probability for the other protocols. These figures show the results obtained varying the number of cooperating sensors for $P = 1000$ and $K = 12, 15$ and 20 respectively. From these three figures it is possible to notice that increasing the key-ring size, the channel exis-

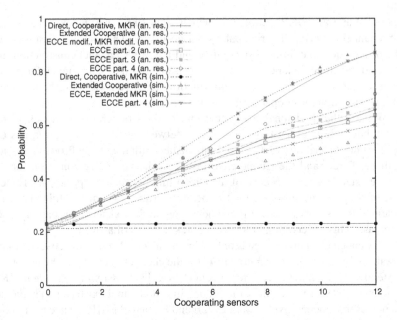

Fig. 2.4 Channel existence: Comparison between analytical and simulation results for $P = 100$, $K = 5$

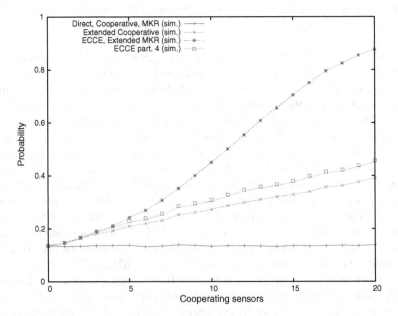

Fig. 2.5 Channel existence: $P = 1000$, $K = 12$

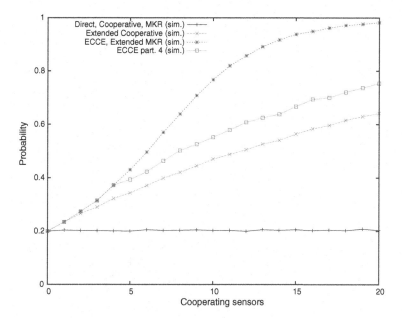

Fig. 2.6 Channel existence: $P = 1000$, $K = 15$

tence probability increases as well. Hence, it is possible to obtain the same existence probability with different key-ring size, varying the number of involved cooperating sensors. The better performance of the ECCE Protocol, compared to the Cooperative one is due to the greater number of possible paths between a and b generated by the ECCE protocol. Indeed, the higher the number of possible paths, the higher the probability of channel existence, as analytically exposed in Sect. 2.5. For instance, in Fig. 2.1, the principals cannot set-up a secure channel via the Cooperative Protocol, while this is possible adopting the ECCE Protocol.

Figure 2.8 shows the existence probability of a channel for the different compared protocols, when $P = 10,000$ and $K = 50$, while $|\mathscr{C}|$ ranges from 0 to 20. We can notice that the curves behaviour is similar to that obtained in Fig. 2.6. This is because, as noticed in Sect. 2.5, the overall channel probability existence strictly depends on the existence probability of a single link.

Furthermore, we inquired the resilience of the established channels. In particular, we have performed our analysis assuming the existence between the two principals of at least the Direct channel, while the cooperating sensors are randomly selected. In Figs. 2.9, 2.10, 2.11, and 2.12 we report on the x axis the key ring size, while on the y axis the number of sensors to corrupt to compromise a channel, considering $P = 10,000$ and 4 and 16 cooperating sensors, respectively.

From Fig. 2.9, with 150 keys stored per sensor, the attacker has to capture about 82 sensors to corrupt a Direct channel; if 4 cooperating sensors are involved, the attacker has to corrupt about 113 sensors to corrupt a Cooperative channel, a little

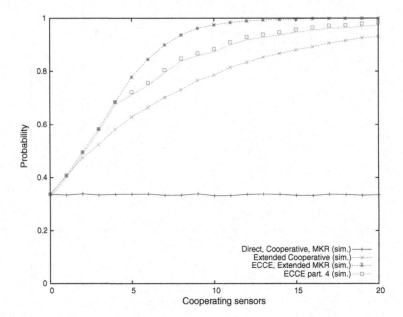

Fig. 2.7 Channel existence: $P = 1000$, $K = 20$

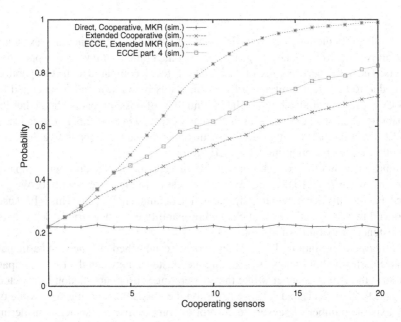

Fig. 2.8 Channel existence: $P = 10,000$, $K = 50$

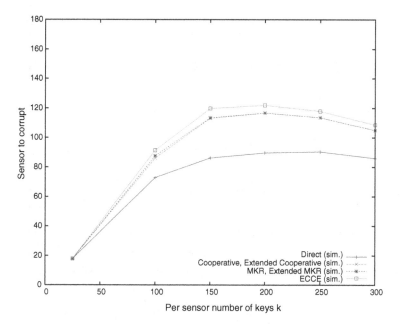

Fig. 2.9 Resiliency: $P = 10,000$, $\mathscr{C} = 4$

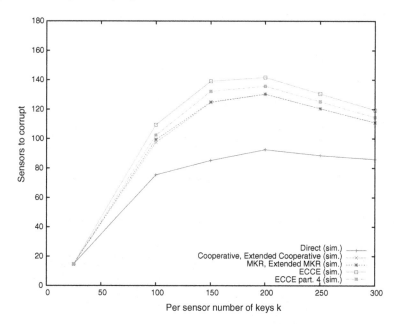

Fig. 2.10 Resiliency: $P = 10,000$, $\mathscr{C} = 8$

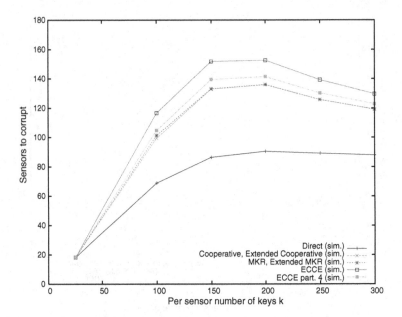

Fig. 2.11 Resiliency: $P = 10,000$, $\mathscr{C} = 12$

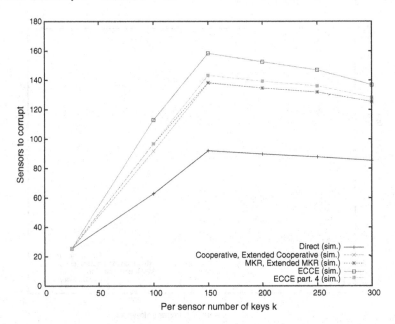

Fig. 2.12 Resiliency: $P = 10,000$, $\mathscr{C} = 16$

more to corrupt the MKR channel, while 120 sensors are required to corrupt an ECCE channel. Increasing the number of cooperating sensors also improves the resilience of these protocols. Indeed, in Fig. 2.12, with 16 cooperating sensors, for a key-ring size of 150, the attacker needs to corrupt about 138 sensors to corrupt the Cooperative, a little more to corrupt the MKR channel, while the attacker is required to compromise about 159 sensors to corrupt an ECCE channel. It is worth noticing that for all the simulated scenarios, the ECCE protocol performs better than the Cooperative protocol. In Figs. 2.9, 2.10, 2.11, and 2.12 the behaviour of the Direct Protocol is the same: The resilience of the Direct channel is not influenced by the number of cooperating sensors. From these figures we can also notice that the resilience to corruption of these protocol increases as the key ring size increases up to a certain value, after that value, the resiliency to corruption decreases. As observed in [85], this is due to the number of keys that the attacker can acquire tampering with a sensor.

Table 2.2 outlines the features of the ECCE Protocol, compared with the others protocols reported in Sect. 2.2. As can be seen, the ECCE Protocol benefits from all the features introduced by the use of cooperation among sensors. This explains why the ECCE Protocol performs better than the other protocols, as shown in the previous figures.

Table 2.2 Features comparison of different protocols

Protocol	Features				
	Involve cooperating sensors	Principals mutual authentication	Usable in the case of no secret shared between principals	Authentication between cooperating sensors	Cooperating sensors that do not share keys with principals helps channel establishment
Direct [74]	No	–	–	–	–
Multipath key reinforcement [37]	Yes	No	No	No	Yes
Cooperative [74]	Yes	Yes	No	No	No
Extended cooperative (this chapter)	Yes	Yes	Yes	No	No
ECCE (this chapter)	Yes	Yes	Yes	Yes	Yes

2.7 Concluding Remarks

In this chapter we presented ECCE, a new cooperative protocol to establish a secure pair-wise communication channel between any pair of sensors in a WSN. The contributions are the following: This protocol does not require cooperating sensors to share a key with both principals for the channel between principals to be established. Also cooperating sensors that do not share any key with any of the two principals can help in the set up of the secure channel; cooperating sensors implement a probabilistic authentication of both principals as well as other cooperators. The probability that a fake principal or a fake cooperator could escape the authentication procedure decreases exponentially with the number of cooperators involved in the protocol; it is possible to trade-off key ring size with the number of cooperating sensors while preserving the same level of security. Note that this feature gives the possibility to have some memory available to store the ECCE keys, that could be used later for further use, hence amortizing the (limited) overhead incurred in the ECCE key set-up; the security provided by the protocol is adaptive with the level of threat in the WSN, on one hand, the higher the security threat, the more cooperators can be involved to enhance the resiliency of the channel; on the other hand a required downgrade on the required level of security can be implemented involving less cooperators, hence improving performances. Finally, in comparison with other protocols, ECCE shows better performances in channel existence and channel resilience even when the number of involved cooperators is small.

While the aim of this chapter is to propose an efficient way to involve cooperating nodes in the pair-wise key establishment, we note that it is also interesting to study how the network density influences the availability of neighbour nodes; a further detailed study on the energy consumed by the protocol would be of interest as well. As for comparing the proposed protocol with the current solutions in the literature, we only considered other probabilistic algorithms: we did not consider deterministic solutions because of the drawbacks described in Sect. 2.2. A broader comparison (e.g. for the channel resilience) of the proposed solution against the deterministic solutions like the one in [22] would also be of interest. We leave these points as future works.

As discussed in Sect. 2.5.2, the main threat to the protocol presented in this chapter comes from the fact that the attacker can capture nodes—acquiring the secret material stored in their memory. If the network were able to know the ID of the captured node it could react somehow. As an example, the pre-deployed keys associated with the captured ID could be revoked—not considered secure anymore. Nodes that do not share anymore a secure channel would stop communicating or will just use a multihop path to communicate or to share a new secret key. Next chapter provides a mechanism to deal with the detection of the node capture.

Chapter 3
Capture Detection

The previous chapter looked at the security of WSNs from the single node perspective. We proposed a new probabilistic protocol for the node authentication and the communication confidentiality. Given that in our solution a secret key can be shared between more than two nodes, the attacker that physically captures a node is potentially able to compromise the confidentiality of a fraction of network communications. Actually, one of the most vexing problems for WSNs security is the node capture attack: An adversary can capture a node from the network eventually acquiring all the cryptographic material stored in it. Further, the captured node can also be reprogrammed by the adversary and re-deployed in the network in order to perform malicious activities.

This chapter addresses the node capture attack in mobile WSNs. In particular, we start from the intuition that mobility, in conjunction with a reduced amount of local cooperation, helps to compute, effectively and with a limited resource usage, the global security properties of the network. Then, we develop this intuition and use it to design a protocol that the network nodes can use to detect the node capture attack. We support our proposal with a wide set of experiments, showing that mobile networks can leverage mobility to compute global security properties, like node capture detection, with a small overhead.

3.1 Introduction

Ad hoc network can be deployed in harsh environments to fulfill law enforcement, search-and-rescue, disaster recovery, and other civil applications. Due to their nature, WSNs are often unattended, hence prone to different kinds of novel attacks. For instance, an adversary could eavesdrop all the network communications. Further, the adversary might capture (i.e., remove) nodes from the network. These nodes can then be re-programmed and deployed within the network area, for instance, to subvert the data aggregation or the decision making process in the network [37]. Also, the adversary could perform a *sybil attack* [159], where a single node illegitimately

© Springer Science+Business Media New York 2016
M. Conti, *Secure Wireless Sensor Networks*, Advances in Information Security 65,
DOI 10.1007/978-1-4939-3460-7_3

claims multiple identities also stolen from previously captured nodes. Another type of attack is the *clone* attack, where the node is first captured, then tampered with, re-programmed, and finally replicated in the network. The former attack can be efficiently addressed with a mechanism based on RSSI [68] or with authentication based on the knowledge of a fixed key set [75], while other solutions have been proposed also for the detection of the clone attack [48, 169, 232].

To think of a foreseeable application for node capture detection, note that, the U.S. Defense Advanced Research Projects Agency (DARPA) initiated a new research program to develop so-called LANdroids [124]: Smart robotic radio relay nodes for battlefield deployment. LANdroid mobile nodes are supposed to be deployed in hostile environment, establish an ad-hoc network, and provide connectivity as well as valuable information for soldiers that would later approach the deployment area. LANdroids might retain valuable information for a long time, until soldiers move close to the network. In the interim, the adversary might attempt to capture one of these nodes. We are not interested in the goals of the capture (that could be, for instance, to re-program the node to infiltrate the network, or simply extracting the information stored in it); but on the open problem of how to detect the node capture that represents, as shown by the above cited examples, a possible first step to jeopardize a WSN. Indeed, an adversary has often to capture a node to tamper with— that is, to compromise its key set, or to reprogram it with malicious code—before being able to launch other more vicious, and may be still unknown, attacks. Node capture is one of the most vexing problems in sensor network security [171]. In fact, it is a very powerful attack and very hard to detect. We believe that any solution to this problem has to meet the following requirements: (i) to detect the node capture as early as possible; (ii) to have a low rate of false positives—nodes that are believed to be captured and thus subject to a revocation process, but that were not actually taken by the adversary; (iii) to introduce a small overhead.

The solutions proposed so far are a long way from being efficient [171]. Also, while naïve centralized solutions can be applied to generic ad-hoc networks, they present drawbacks like single point of failure and non uniform energy consumption. These drawbacks do not make them appealing for sensor networks. Moreover, these networks often operates without the support of a base station. Efficient and distributed solutions to the node capture attack are of particular interest in this context.

To the best of our knowledge, there are no distributed solutions for the problem of detecting the node capture attack in WSN. Following a new interesting research thread that focuses on leveraging mobility to enforce security properties for wireless sensor networks [178, 218], we propose a capture detection framework exploiting node mobility. We show that this approach can provide better performance compared to traditional solutions. Also, we show that using node cooperation in conjunction with node mobility can still improve the capture detection performance within specific network requirements.

The contribution of this chapter is to provide a proof of concept: It is possible to leverage the emergent properties of mobile sensor networks via node mobility and node cooperation to design a node capture detection protocol. To this aim, we use the Random Waypoint Mobility Model (RWM) [25], an ideal mobility model that is simple and general enough (at least for some application scenarios) to explore our ideas. Furthermore, the result on any particular mobility model should depend not only from the model but also from the network setting, as pointed out in [203] for the delay-capacity trade-off. Indeed, providing specific settings and evaluations for other models is out of the scope of this work.

Our solutions are based on the simple observation that if node a will not *re-meet* node b within a period λ, than it is possible that node b has been captured. We will build upon this intuition to provide a protocol that makes use of local cooperation and mobility to locally decide, with a certain probability, whether a node has been captured or not. Our proposed solutions do not rely on any specific routing protocol: We resort to one-hop communications and to a sparing use of a message broadcasting primitive. This distinguished feature helps keep our protocol simple, efficient, and practically deployable, avoiding the use of sophisticated routing that can introduce complexity and overhead in the mobile setting. Furthermore, our experimental results demonstrate the effectiveness and the efficiency of our proposal. For instance, for a given energy budget, while the benchmark requires about 4,000 s to detect node capture, our proposal requires less than 2,000 s. Moreover, our later work on node capture attack is presented in [45].

Organization

The chapter is organized as follows. Section 3.2 presents the related work in this area. Section 3.3 introduces the motivation and the framework of our proposal based on simple sensor network capabilities like node mobility and message broadcasting. Our specific proposal, the CMC Protocol, is then presented in Sect. 3.4, while in Sect. 3.5 we discuss the simulation results that give a qualitative idea of how mobility and node cooperation can be exploited in order to decrease the node capture detection time. Finally, Sect. 3.6 reports some concluding remarks.

3.2 Related Work and Background

Mobility as a mean to enforce security in mobile networks has been considered in [218]. Further, mobility has been considered in the context of routing [102] and of network property optimization [147]. In particular, [102] exploits node mobility in order to disseminate information about destination location without incurring any communication overhead. In [147] the sink mobility is leveraged to optimize the energy consumption of the whole network. A mobility-based solution for detecting the sybil attack has been presented in [178]. Finally, note that a few solutions exist for node failure detection in ad hoc networks [110, 114, 115, 184]. However, such solutions assume a static network, missing a fundamental component of our scenario, as shown in the following.

In this chapter we use node mobility to cope with the node capture attack. As described in the following section, we specifically rely on the meeting frequencies between honest nodes to gather information about the absence of captured nodes. A property similar to that of node "re-meeting" has been already considered in [64]. However, in [64], the authors investigate the time needed for a node to meet (for the first time) a fixed number of other nodes. This analysis is then used together with node mobility to achieve non-interactive recovery of missed messages. To the best of our knowledge no distributed solution exploiting node mobility has been proposed to detect the node capture attack in mobile ad-hoc and sensor networks.

We published [53] a short contribution on the possibility to leverage network mobility for node capture detection. In particular, in [53] we presented the main intuition and a first basic solution in order to understand rationales of this type of approach. However, while the results given in [53] are encouraging, the specific solution proposed requires an high overhead to bound the number of false positive (wrongly revoked nodes). Note that, without this bounding mechanism the number of false positive would be unacceptable. Furthermore, in [53] we did not study the feasibility of the new approach compared with other ones. In the present work, we leverage the intuition proposed in [53], that is the "re-meeting" time between nodes, to design two brand new efficient protocols. In particular, we introduce a presence-proving mechanism used by allegedly captured nodes to show their actual presence in the network (that is, bounding the number of false positive). Further, we introduce a benchmark solution in order to quantify the quality of the proposed solutions. The proposed solutions are compared between them and with the benchmark. In particular, to have a fair comparison we observed the detection time provided by the different protocols using the same energy budget. The result of our study confirms the feasibility of the approach sketched in [53]. Furthermore, it proves that, within certain scenarios of node mobility, the proposed solutions provide a sensitive improvement over other possible approaches, such as the one based on the classical message exchange.

Node mobility and node cooperation in a mobile ad hoc setting has been considered already in Disruption Tolerant Networks (DTNs) [65, 209]. However, such a message passing paradigm has not been used, so far, to support security. We leverage the concept introduced with DTN to cooperatively control the presence of a network node. In this chapter we use one of the most common mobility patterns in the literature, the Random Waypoint Mobility Model [25]. In this model, it is assumed that each node in the network acts independently: It selects a geographic destination in the deployment area (the *way-point*), it selects a speed uniformly at random in a given interval $[s_{min}, s_{max}]$, where s_{min} and s_{max} stand for minimum and maximum speed respectively.Then it moves towards the destination on a straight route at the selected speed. When at the way-point, it waits for some time, again selected uniformly at random from a given interval, and then the node repeats the process by choosing the next way-point. Some researchers have shown some problems related to this mobility model. One of the problems is that the average speed of the network tends to decrease during the life of the network itself and, if the minimum speed that can be selected by the nodes is zero, then average speed of the system converges to zero [237]. In the same paper it is suggested to set the minimum speed to a value strictly greater than zero. In this case, the average speed of the system continue decreasing, but it converges to a non-zero asymptotic value. Other problems related to spatial node distribution have been considered by different authors [123, 237]. In the analysis presented in [102] "human speeds" are claimed to be a reasonable practical choice for mobile nodes.

Note that the RWM might not be the best model to capture a "realistic" mobility scenario, as highlighted in [203]; however, the results achieved in this chapter are meaningful as they are a proof of concept that mobility can be leveraged to enforce security properties and also as the provided protocols could be used in, and adapted to, more realistic mobility models. In our proposed approach every node maintains its own clock. However, we require that clocks among nodes are loosely synchronized. Note that there are a few solutions proposed in the literature to provide loose time synchronization, like [211] for wireless sensor networks; therefore in the following we will assume that skew and drift errors are negligible.

The problem addressed in this chapter can be seen as a topology control problem, where locations, and ranges of nodes need to be securely obtained prior to make any statements about nodes' presence in the network. An overview of the issues related to the communication versus physical neighbourhood can be found in [168], while an example of the use of distance-bounding for topology control can be found in [215]. We observe that the solution presented in this chapter can be further extended using these approaches.

In our proposal we also need to take into consideration the cost of broadcasting a message to all the nodes in the network. In [229] a classification of the different solutions for broadcasting scheme is provided: (i) Simple Flooding; (ii) probabilistic-based schemes; (iii) area based schemes that assumes location awareness; (iv) neighbour knowledge schemes that assumes knowledge of two hop neighbourhood.

Analyzing or comparing broadcasting cost is out of the scope of this chapter. However, for a better comparison of the solutions proposed in this chapter, we need to fix a broadcast cost that will be expressed in terms of unicast messages. In fact, the overhead associated to the broadcasting varies with different network parameters (for instance, node density and communication radius). A deeper analysis on the overhead generated for different broadcasting protocols is presented in [163]. Also, note that probabilistic-based and neighbour-based protocols require a big overhead for a mobile network in order to know the network topology and neighbourhood respectively. Furthermore, the same argument can be considered for the localization protocol that is used in the area-based schemes. In the following, to embrace the more general case, we assume that nodes are not equipped with localization devices, like GPS.

Finally, note that a message could be received more than once, for instance because the receiver is in the transmission range of different relay nodes. However, in the following we assume that a broadcasted message is received (then counted) only once for each node. This reflect the usual practice to switch off the radio transceiver as soon as a node is aware it is receiving a copy of the same message just broadcasted. Indeed, the node will then go to sleep over the rest of the message transmission, thus consuming negligible power. Therefore, we can consider negligible the overall energy consumption associated to this operation. A similar assumption is used for example in [163].

3.3 Node Capture Detection Through Mobility and Cooperation

The aim of a capture detection protocol is to detect as soon as possible that a node has been removed from the network. In the following we also refer to this event as a node capture. The protocol should be able to identify which is the captured node, so that its ID can be revoked from the network. Revocation is a fundamental feature—if the adversary reintroduces the captured (and possibly reprogrammed) node in the network, the node should not be able to take part to the network operations.

In the following we first describe a simple distributed solution that does not exploit neither mobility nor cooperation among nodes; we use this solution as a benchmark to compare with our proposal. Then, we introduce the rationals we leverage to develop our protocol for node capture detection, detailed in next section.

3.3.1 Benchmark Solution

To the best of our knowledge, no efficient and distributed solution exploiting mobility was proposed so far to cope with the node capture detection problem in Mobile Ad Hoc Network. However a naïve solution exploiting node communication capabilities can be easily figured out. We first describe this solution assuming the presence of a base station (BS); then, we will show how to relax this assumption. In the BS-based solution, each node periodically sends a message to the BS carrying some evidence of its own presence. In this way the base station can witness for the presence of the claiming nodes. If a node does not send the claim of its presence to the BS within a given time range, the base station will revoke the corresponding node ID from the network (for instance, flooding the network with a revocation message). To remove the centralization point given by the presence of the BS, we require each node to notify its presence to any other node in the network. To achieve this goal, every t seconds a node sends a claim message advertising of its presence all the network nodes through a broadcasted message. A node receiving this claim would restart a time-out set to $t + \sigma$ where σ accounts for network propagation delay. Should the presence claim not be received before the time-out elapses, the revocation procedure would be triggered. However, note that if a node is required to store the ID of any other node as well as the receiving time of the received claim message, $O(N)$ memory locations would be needed in every node, where N is the number of nodes in the WSN. To reduce the memory requirement on node, it is possible to assume that the presence in the network of each node is tracked by a small subset of the nodes of the network. Hence, if a node is absent from the network for more than t seconds, its absence can still be detected by a set of nodes.

3.3.2 Our Approach

Our approach is based on the intuition that leveraging node mobility and cooperation helps node capture detection. We start from the following observation: If node a has listened to a transmission originated by node b, at time t, we will say that a *meeting* occurred. Now, nodes a and b are mobile, so they will leave the communication range of each other after some time. However, we expect these two nodes to re-meet again within a certain interval of time, or at least within a certain time interval with a certain probability. The solution can also be thought of an exploitation of the opportunistic communication concept [209], like contact-based message delivery, to wireless sensor network security. In [53] the authors investigated how mobility can be exploited for detecting a node capture and investigated the feasibility of mobility based solutions. As a starting point, we analysed the re-meeting probability through network simulation: The results comply with previous studies on delay in mobile ad hoc networks [203]. In Fig. 3.1 we report on the simulation results on the probability that two nodes that had a meeting, would not have a meeting again after x seconds. This probability has been evaluated for different values of the communication radius.

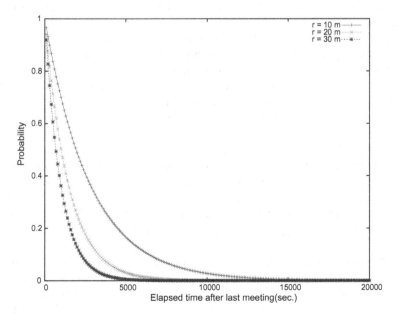

Fig. 3.1 Non cooperative approach. Probability for two nodes not to re-meet again: $N = 100$, $s_{min} = 5\,\text{m/s}$, $s_{max} = 15\,\text{m/s}$

In particular, we assume that the nodes are randomly deployed in a square area of 1,000 m × 1,000 m and that they move according to the random way-point mobility model. While the x-axis indicates the time after the last meeting, the y-axis indicates the probability that the two nodes have not re-met yet. For example, assume that node a meets node b at time t, then the probability that these two nodes have not met again after 5,000 s is very close to 0 (for a sensing radius r = 30).

In the following section we propose a protocol that leverages node mobility to enhance node capture detection probability.

3.3.3 Assumptions and Notation

In the remaining of the chapter we assume a "smart" attacker model: It knows the detection protocol implemented in the network. This implies, for the Benchmark solution, that a node a is captured just after node a has broadcasted its presence claim message. The assumption at the base of our protocol is that if a node has been absent from the network for a given interval time (that is no one can prove its presence in that interval) the node has been captured. Indeed, we could incur in wrong revocation if the node is actually not captured but, for example, only disconnected for that considered time interval. It is worth noticing that, also if a node is temporarily disconnected, a DTN-like routing mechanism [28] can be used to deliver a message to that node with some delay. For the aim of our protocol, we do not explicitly consider that interval time. We assume that it is comprised within the interval time a node has to prove its presence, once accused to be captured.

In the following we define a *false positive alarm* as an alarm raised for a node that is actually present. One or more false positive alarms can imply a *false positive detection*, that corresponds to the revocation of a not captured node. Further, we refer to a *false negative detection* as a captured node not actually revoked. Note that a node that is temporarily disconnected, but not captured, could be revoked from the network as well. Another issue is Denial of Service (DoS). Indeed, since alarm are flooded in the network (as it will be clear in the following), it could be possible for a corrupted node to trigger false alarms so as to generate a DoS. This issue is out of the scope of this chapter, however, for the sake of completeness, we sketch in the following a possible solution. The impact of false positives can be mitigated noting that it could be possible, once the recovery mechanism detects a false alarm, to associate a failure tally to the node that raised the false alarm. If the tally exceeds a certain threshold, the appropriate action to isolate the misbehaving node could be take.

Further, we assume the existence of a failure-free node broadcasting mechanism [146]; and, finally, we point out that addressing node-to-node secure communications, addressing confidentiality, integrity, privacy, and authentication are out of the scope of this chapter. However, note that a few solutions explicitly addressing these points can be found in literature [75, 183, 210].

Table 3.1 resumes the intervals time notation used in this chapter.

Table 3.1 Time-related notation

Symbol	Meaning
σ	Message propagation delay
λ	Alarm time used in CMC (our proposal)
δ	Time available to the allegedly captured node to prove its presence

3.4 The Protocol

In this section we describe our proposal for a node Capture detection protocol that leverages Mobility and Cooperation (CMC Protocol). Basically, each node a is given the task of witnessing for the presence of a specific set T_a of other nodes (we will say that a is *tracking* nodes in T_a). For each node $b \in T_a$ that a gets into the communication range of, a sets the corresponding meeting time to the value of its internal clock and starts the corresponding time-out, that would expire after λ seconds. The meeting nodes can also cooperate, exchanging information on the meeting time of nodes of interests—that is, nodes that are tracked by both a and b. Note that node cooperation is an option that can be enabled or disabled in our protocol. If the time-out expires (that is, a and b did not re-meet within λ seconds), the network is flooded with an alarm triggered by node a. If node b does not prove its presence within δ seconds after the broadcasted alarm is flooded, every node in the network will revoke node b. The detailed description of the CMC protocol follows.

3.4.1 Protocol Description

The CMC protocol is event based; in particular it is executed when:

- Node a meets node b: This event triggers node a and node b to execute CMC_MEETING (ID_b, *false*, $-$) and CMC_MEETING (ID_a, *false*, $-$) respectively, if the cooperation parameter is set to false. Otherwise, node a executes CMC_MEETING (ID_b, *true*, $-$) and node b executes CMC_MEETING (ID_a, *true*, $-$). The function CMC_MEETING is also used in the cooperative scenario as a *virtual* meeting in order to update node presence information.
- The time-out related to node ID_x expires on node a: Procedure CMC_TIMEOUT (ID_x) is executed by node a.
- Node a eavesdrops a message m: Node a executes the procedure CMC_RECEIVE(M).

Algorithms 4, 5, and 6 show the corresponding pseudo-code.

The procedure CMC_MEETING, shown in Algorithm 4, is executed by both nodes involved in a meeting. In the case of a real meeting the time is not specified, then the current node time t_a is used. However, when the procedure is invoked as a *virtual* meeting a reference time (t_x) is also considered (lines 2, 3 and 4). When node a meets node b, node a checks if it is supposed to trace node b (that is if $b \in T_a$). This is done by invoking the function TRACE (line 5). This function takes in input two node IDs, and provides a result pseudo-uniformly distributed in $\left[1..\left\lceil\frac{N}{|T|}\right\rceil\right]$; where N is the size of the wireless sensor network and $|T|$ is the number of nodes tracked by each node. Node b is to be tracked if and only if the result is one. A simple and efficient implementation of the function TRACE can be found in [74], where it has been used in the context of pairwise key establishment. Assume now that $b \in T_a$, then a further check on node b is performed (line 6). Indeed, node b could be already revoked. Hence, each node stores a Revocation Table (RT_a) that lists the revoked nodes. If both previous tests (line 5 and line 6) succeed, then a calls the function Update that updates the information about the last meeting with node b (line 7). For example, if node a meets b at a given time t_a, the function Update sets the information $\langle ID_b, t_a \rangle$ in the CT_a (a Check Table stored in node a memory). Node a uses a Time-out Table TT_a to store and signal the following time-outs:

- ALARM time-out, that is triggered after λ seconds are elapsed without re-meeting node b.
- REVOKE time-out, that is triggered after δ seconds are elapsed from receiving/triggering a node revocation for node b—assuming that in these δ seconds no presence claim from b are received.

Then, for each meeting with not-revoked nodes in T_a, node a removes any previous time-out for the met node and sets a new ALARM time-out for that node (line 8). Note that, both the update functions (lines 7 and 8) do not perform any operation if the time argument t_x is lower than the currently stored meeting time for the node ID_x: This could happen in the case of a *virtual* meeting.

If the cooperation option is set ($COOP\_opt = true$ in line 11) also the following steps are performed. For each not revoked node x traced by both node a and b (lines 12, 13, and 14), node a sends a CLAIM message to b carrying the meeting time between a and x. Each CLAIM message has the following format: $\langle ID_a, CLAIM, ID_x, \text{elapsed time} \rangle$, where ID_a is the sender of the claim message, CLAIM is the message type, ID_x is the ID of node x the claim is related to, and the last parameter indicates the meeting time between a and x. Another message type is ALARM, described in the following.

CMC_TIMEOUT (Algorithm 5) is triggered when a time-out expires. If on node a an ALARM time-out expires for node ID_b, this means that node a did not meet node ID_b for a time λ. Then node a floods the network with an alarm (Algorithm 5, line 3) and a new REVOKE time-out for node b is set. Each ALARM message has the following format: $\langle ID_a, ALARM, ID_b \rangle$, where ID_a is the sender of the claim message, ALARM notifies the message type, and ID_b is the ID of node b the alarm is related to. When a REVOKE time out expires this means that, after δ seconds elapsed from the alarm triggering, no evidence of the presence in the network of the suspected captured node appeared. In this latter case a node revocation procedure for node b is invoked by node a.

CMC_RECEIVE (Algorithm 6) is invoked when a message MSG is received. The fields of the message are assigned to local variables (line 2) and the type of the

---

Input : ID_a : ID of the executing node. ID_b : ID of the met node. t_a : Current time of node a. CT_a : Check Table stored in node a memory. RT_a : Revoked nodes table stored in node a memory. TT_a : Time out table stored in node a memory. λ : Alarm time. δ : Time for the accused node to prove its presence. $COOP\_opt$: Boolean variable for cooperation option.

```
 1 begin
 2 |  if NotSpecified(t_x) then
 3 |  |  t_x = t_a;
 4 |  end
 5 |  if Trace(ID_a,ID_b)=1 then
 6 |  |  if IsNotRevoked(RT_a,ID_b) then
 7 |  |  |  Update(CT_a,⟨ID_b,t_x⟩) ;
 8 |  |  |  UpdateTimeOut(TT_a,
        ⟨ID_b,t_x + λ, ALARM⟩) ;
 9 |  |  end
10 |  end
11 |  if COOP_opt = true then
12 |  |  foreach ⟨ID_x, t_x⟩ ∈ CT_a do
13 |  |  |  if IsNotRevoked(RT_a,ID_b) then
14 |  |  |  |  if Trace(ID_b, ID_x)=1 then
15 |  |  |  |  |  ⟨t_old⟩ ←LookUp(CT_a,ID_x) ;
16 |  |  |  |  |  ⟨ID_a, CLAIM, ID_x, t_old⟩ → b ;
17 |  |  |  |  end
18 |  |  |  end
19 |  |  end
20 |  end
21 end
```

Algorithm 4: CMC_MEETING(ID_x, $COOP\_opt$, t_x).

Input : ID_a : ID of the executing node. ID_b : ID of the node which time-out is expired. t_a : Current time of node a. RT_a : Revoked nodes table stored in node a memory. TT_a : Time out table stored in node a memory. δ : Time for the accused node to prove its presence.

```
1 begin
2 |  if TimeOutKind(ALARM) then
3 |  |  Flooding(⟨ID_a, ALARM, ID_b⟩) ;
4 |  |  UpdateTimeOut(TT_a,
        ⟨ID_b, t_a + δ, REVOKE⟩) ;
5 |  else
6 |  |  RevokeNode(RT_a,ID_x)
7 |  end
8 end
```

Algorithm 5: CMC_TIMEOUT(ID_x).

message is checked (line 3). Assume the message is of type ALARM: The executing node checks if the alarm is related to itself (line 4).

If the latter test fails, a further check is performed: The node checks whether the node ID_x is not already revoked (line 5). If the check succeeds, a REVOKE time-out is set through an UpdateTimeOut procedure. Note that, should a REVOKE time-out for node b already be in place, this procedure does not override the existing REVOKE time-out and simply returns. If the ALARM is related to the executing node itself (test performed at line 4 fails) node a will flood the network with a presence CLAIM message (line 9). This measure prevents *false positive detection*—that is, the revocation of nodes that are active in the network.

If the received message is of type CLAIM, this means that a node that was the target of an ALARM message is proving its presence; this message triggers a *virtual* meeting between a and the wrongly accused nodes (line 13). The overall result is that node a disables the REVOKE time-out for that node while restarting the ALARM time-out for the same node. These activities are also triggered when the *COOP_opt* is set (in fact, a CLAIM message is also sent in line 16, Algorithm 4). The objective of this invocation is to update the information on traced nodes via an information exchange with the met nodes.

Finally, when a receives a message issued by node b that is not originated within the protocol (for instance, it can be originated by the application layer), this message can be interpreted by the protocol as an evidence of the presence of node b. Therefore, this can be interpreted as a special case of a node meeting, and the appropriate actions are triggered (line 15).

3.5 Simulations and Discussion

We performed simulations using a self-developed discrete event simulator. As for the energy model, we adopted the one proposed in [222]. To plot each point in the following graphs (as well as for Fig. 3.1), we performed a set of experiments and reported the averaged results; the number of experiments has been set to achieve a confidence interval of 98 %.

The comparison on the detection time between our protocol and the benchmark has been performed considering the energy cost. In particular, the energy cost has been expressed as a frequency of network flooding, as explained later.

3.5.1 Node Re-Meeting

In order to better understand how mobility and cooperation can speed up the capture detection process we performed a first set of simulations to assess the frequency of node-to-node meetings. We considered a network of $N = 100$ nodes randomly deployed over a square area of $1,000\,\text{m} \times 1,000\,\text{m}$. We used the random

Input : ID_a : ID of the executing node. t_a : Current time of node a. MSG : Received
message. RT_a : Revocation Table stored in node a memory. δ : Time for the accused
node to prove its presence.

```
 1 begin
 2 │   ⟨ID_b, msg_type, ID_x, t_x⟩ ← MSG ;
 3 │   if (msg_type = ALARM) then
 4 │   │   if (ID_x ≠ ID_a) then
 5 │   │   │   if IsNotRevoked(RT_a,ID_x) then
 6 │   │   │   │   UpdateTimeOut(TT_a,
 │   │   │   │   ⟨ID_b, t_a + δ, REVOKE⟩) ;
 7 │   │   end
 8 │   │   else
 9 │   │   │   Flooding(⟨ID_a, CLAIM, −, −⟩) ;
10 │   │   end
11 │   end
12 │   if (msg_type = CLAIM) then
13 │   │   CMC_Meeting(ID_x, false, t_x) ;
14 │   end
15 │   CMC_Meeting(ID_b, false, −) ;
16 end
```

Algorithm 6: CMC_RECEIVE(*MSG*).

waypoint mobility model as the node mobility pattern. In particular, in our simulations we set the value for the minimum node speed greater than zero—this is a way to solve the decreasing average node speed problem of the random waypoint mobility model [237].

The experiment was set in this way: We choose two nodes a and b, when they meet, we set time at $t = 0$ and continued following these nodes thorough their network evolution to experimentally determine how long it takes for these two nodes to meet again, in both the non-cooperative and in the cooperative case.

Crucially, in the cooperative scenario, if node c meets node a and sends to it all the information c received during its last meeting with node b, this is also a meeting between a and b.

We performed the simulation for different values of sensing radius and average node speed both for the non-cooperative and the cooperative scenario. The results are shown in Figs. 3.2 and 3.3. The experiments support the following, simple intuitions: Node cooperation increases the meeting probability; the higher is the sensing radius, the higher is the meeting probability; the higher is the average node speed, the higher is the meeting probability.

We used these results also to propose a reasonable value for the variable λ to be used in the implementation of our proposal, for both the cooperative and non-cooperative case.

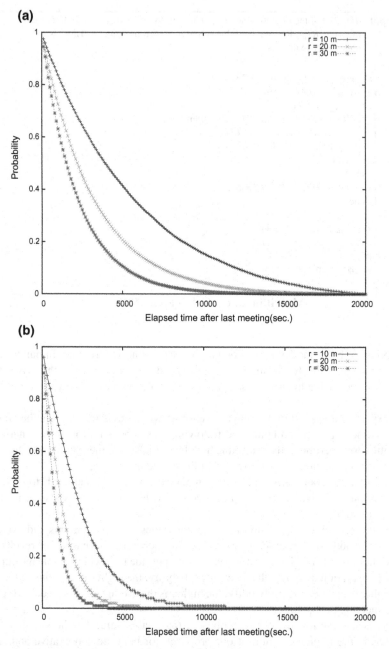

Fig. 3.2 Probability for two nodes not to re-meet: $N = 100$, $s_{avg} = 5$ m/s **a** Without node cooperation, $s_{avg} = 5$ m/s. **b** With node cooperation, $s_{avg} = 5$ m/s

(a)

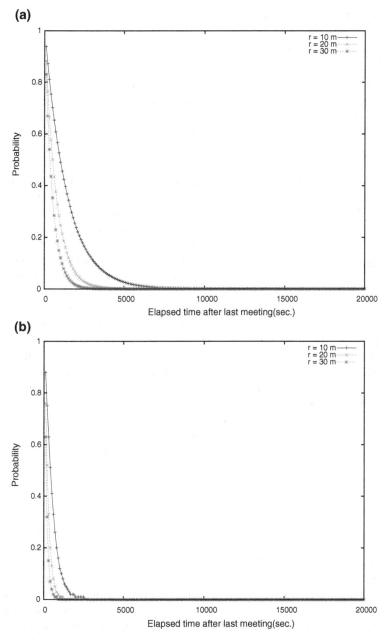

Fig. 3.3 Probability for two nodes not to re-meet: $N = 100$, $s_{avg} = 20$ m/s **a** Without node cooperation, $s_{avg} = 20$ m/s. **b** With node cooperation, $s_{avg} = 20$ m/s

3.5.2 Experimental Results

Energy-driven comparison

One of the key issues in sensor network is the energy consumption. Hence, we compared our proposal with the benchmark focusing on energy consumption. To provide an evaluation of our protocols in a manner that is device-independent, we chose to express the energy consumption in terms of generated messages. As for the energy devoted to computation, we considered that cost negligible, as in [222].

The main communication cost of both our protocol and the benchmark are the floodings. The benchmark uses the flooding as a presence claim message while our protocol uses the flooding for both alarm broadcast and alarm-triggered presence notification; the latter flooding occurs when a node that has been erroneously advertised as possibly compromised, sends (floods) a claim of its actual presence. To simplify our discussion, we assume that a network flooding corresponds to send and to receive a message by each network node. This is not always the case; actually, the load for broadcasting varies with different network parameters and the specific broadcasting protocol used [163]. However, this approximation is good enough to achieve our goal, that is to show the qualitative improvement of our solution over the benchmark. To better appreciate the comparison with the benchmark—where a flooding occurs every time interval—in the following graphs we report on the x-axis the time interval between two subsequent flooding, instead of the flooding frequency. Note that once the flooding interval is fixed, also the amount of available energy (messages) is fixed, and we can plot the performance of our protocol when using the same amount of energy (that is, the same amount of messages).

In our simulation, we analyze how increasing the energy overhead affects the detection time. In other words, we fix the energy overhead at the same level for both protocols under evaluation, and measure which protocol achieves the best detection time.

Performance

To compare the performance of the proposed solution with the benchmark presented in Sect. 3.3.1, we implemented our protocol. In the following we fix a sensing radius of $r = 20$ m. Since sensor nodes have often strict memory constraints (as for example in sensor network), in our simulations we assume that each node traces a small number of other nodes. In fact, as a result of the pseudo-random function TRACE (Algorithm 4, line 2) each node traces exactly 5 other network nodes. For the cooperative scenario when two nodes a and b meet, they exchange the information concerning the nodes tracked by both a and b; we assume that this information can be contained in one message. Indeed, the number of shared traced nodes can be up to 5 (number of nodes traced by each node), but in practice it turns out to be much smaller, on average (0.25 in our setting). We simulated our protocol with and without node cooperation, varying the alarm time from 250 to 8000 s and the average node speed from 5 to

20 m/s. Figures 3.4a, b show the results of the simulation of our protocol without and with cooperation, respectively.

Figure 3.4a shows the results when cooperation is switched off, for the two protocols and different speeds. On the x-axis, we fix the flooding interval for the benchmark protocol. In this way, the detection time is also fixed for the benchmark and it does not change when changing the speed. The quality of the detection for the benchmark is just linear: By doubling the flooding interval also the detection time doubles, while the energy cost halves. Figure 3.4a confirms our intuition: Mobility with local cooperation can help computing global properties cheaply.

In this simulation scenario, for a reasonable speed of nodes, our protocol outperforms the benchmark. Take, as an example, a flooding interval of 50 s. From Figure 3.4a, we can see that the detection time of the benchmark protocol is 5,000 s. The performance of our protocol depends on the average speed of the system. If the average speed of the system is slow, for example 5 m/s, then the detection time is more than 6,000 s. However, if the network nodes move faster, than our solution improves over the benchmark. For instance, when the average speed is 20 m/s, the detection time is as low as 1,600 s, much faster than the benchmark. From this experiment, it is also clear that the performance of our protocol depends on the average speed in the network: The faster the better. While the benchmark is an excellent solution for slow networks, for example where nodes are carried by humans walking, our solution is the best for faster networks, and it is always the best when the energy overhead must be low. Now, we will switch cooperation on, and see that the performance of our protocol increases considerably, even though with some drawbacks when the energy budget is small.

Figure 3.4b describes the performance of our protocol when using cooperation. When the network flooding frequency is high, that is, network flooding interval is small, cooperation is very effective. Further, with cooperation the performance of our protocol improves as the average speed of the nodes increases. In this case, our protocol is better than the benchmark even when starting from very high flooding frequency—that is, starting from systems that are very fast in detecting the node capture attack and that, consequently, have very high energy requirements. What is less intuitive, is that cooperation is not useful when we move to more energy-saving systems. Take, as an example, a network where the average speed is 15 m/s. Our protocol is better than the benchmark whenever the design goal is to have a network with more energy available and to achieve a small detection time, that is, in Fig. 3.4b, whenever the flooding interval is smaller than 38 s. However, when considering a network with more stringent energy requirements, for example when the flooding interval is 50 s, than it is simply not possible to reach such low energy costs by using cooperation. Cooperation has a cost, that is higher when the network is faster—indeed, in a faster network the nodes meet more frequently, and thus cooperation is higher. In this case, the correct design guideline is to use our protocol with cooperation, if the objective is to have a system that is fast in detecting the node capture attack, though using more energy—in particular, in our example until a flooding interval of 38 s—and then to switch cooperation off, to get a cheaper protocol, that can be used when the flooding interval can be larger.

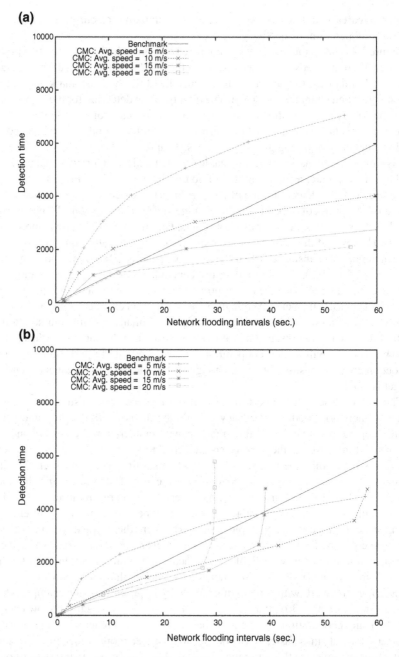

Fig. 3.4 CMC detection time: $N = 100$, $r = 20\,m$ **a** Without node cooperation. **b** Using node cooperation

As described by Fig. 3.4b, the limits of cooperation appear sooner in faster networks. This is intuitive, cooperation is more costly when nodes meet more often, and so the trade-off moves towards non-cooperation earlier. The implications of using mobility and local communications to compute global properties are not trivial. If the network is fast enough, it is always better to use protocols like the one we propose rather than using static approaches like the benchmark. However, node cooperation flavoured techniques, that appears to be effective in any case, have the result of making the information in the network spread faster, but at a cost.

3.5.3 Massive Attacks

In order to investigate the behavior of our protocol under a massive attack we simulated the capture of 10 % of the network nodes (10 out of 100) at the same time. We fixed the average speed at 15 m/s. Simulation results are shown in Figs. 3.5a, b for the non cooperative and cooperative scenario, respectively. For both cases the figures show the result for one captured node and 10 captured nodes in a network of 100 nodes. From both figures we can see that all the protocols, both the benchmark and our solution, with or without cooperation, are robust against massive attacks. Indeed, the small differences in performance do not justify a change in the defense strategy but for small intervals.

3.5.4 Other Mobility Patterns

We stress once again that the aim of this chapter is to give a proof of concept that both node mobility and node cooperation can help thwarting the node capture. Hence, to abstract from mobility details we choose to use the Random Waypoint Mobility Model. Mobility models based on randomly moving nodes may, for example, provide useful analytical approximations to the motion of vehicles that operate in dispatch mode or delivery mode [14]. It is important to note that the results obtained in this chapter are not directly applicable to others scenario-inspired mobility models [203]; for instance, while inter-meeting time follows an exponential distribution under the RWM, inter-meeting time is shown to be better approximated by a power-law distribution in some scenarios [34, 203]. However, it is also interesting to note that our solution allows the network to let autonomously emerge the sub-groups of nodes that meet with higher frequency (communities). In fact, this can be done leveraging the false positive alarm: If node a sends a high number of false alarms (further revoked by the accused node) related to node b this implies that a actually does not meet with b with "high" frequency. This information can be interpreted as if a and b do not belong to the same community.

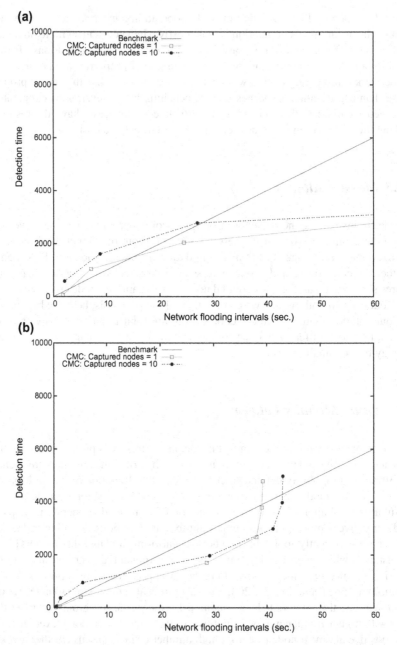

Fig. 3.5 CMC detection time under massive attack: $N = 100$, $r = 20\,m$, $s_{avg} = 15\,m/s$ **a** Without node cooperation. **b** Using node cooperation

3.6 Concluding Remarks

In this chapter we have proposed, to the best of our knowledge, the first distributed solution to a major security threat in mobile WSN: The node capture attack. Our solution is based on the intuition that node mobility, together with local node cooperation, can be exploited to design security protocols that are extremely effective and energy-efficient. These protocols make use, in a non trivial way, of the information flow guaranteed by node mobility. We have also developed a technique that, increasing the level of cooperation among nodes, makes global information flow faster in the network, even if at a cost in terms of energy. The experiments clearly show that these ideas deliver effective and efficient protocols, and that it is possible to find the critical speed that is necessary to induce enough information flow to make these new protocols outperform traditional ones, designed for static networks. We observe that, while the solution provided in this paper has been presented in the context of WSNs, it can be actually extended for a more general type of mobile network: Mobile Ad Hoc Network (MANET). Furthermore, we observe that the approach proposed in this chapter could be improved discovering more stringent conditions to detect an attack; for instance, taking defensive measure for a potential node capture attack subsequently to the detection of a network intrusion would reduce the cost of the defensive mechanism. This solution would apply to the defensive mechanism described in this chapter too. Finally, we believe that the ideas and protocols introduced in this chapter, even if specifically suited to address a major security threat, could be also adopted in other scenario to support other emergent properties as well.

As observed in this chapter, an early capture detection means an higher energy consumption. As a result, because of the energy constraints of WSNs, the detection time cannot be decreased under a given threshold. This make it worth facing the threats coming from a node capture that is not detected. Next chapter specifically addresses one of this threats: The clone attack.

Chapter 4
Clone Detection

The previous chapter considered the node capture problem. The capture of a node affects all the network: All the nodes should be aware that the corresponding node's ID is untrusted from the moment of the capture. We proposed a new approach for the node capture detection. However, the capture detection comes with a cost (in terms of energy consumption) that is inversely proportional to the detection time and should not be prohibitive for the network. As a result, the detection time cannot be decreased under a given threshold. If the node capture is undetected, the adversary can first re-program the captured node and then clone it in a large number of clones, easily taking over the network. Then, the detection of node clone attacks in a wireless sensor network is also a fundamental problem. A few distributed solutions to this problem have been recently proposed in literature. However, these solutions are not satisfactory. First, they are energy and memory demanding: A serious drawback for any protocol to be used in a resource constrained environment like a sensor network. Further, they are vulnerable to specific adversary models introduced in this chapter.

The contributions of this chapter are threefold. First, we analyze the desirable properties of a distributed mechanism for the detection of node clone attacks. Second, we show that the known solutions for this problem do not completely meet our requirements. Third, we propose a new self-healing, randomized, efficient, and distributed protocol (RED) for the network autonomous detection of node clone attacks and we show that it is completely satisfactory with respect to the requirements. Extensive simulations also show that our protocol is highly efficient in communication, memory, and computation, that it sets out an improved attack detection probability on the best solutions in the literature, and that it is resistant to the new kind of attacks we introduce in this chapter, while other solutions are not.

© Springer Science+Business Media New York 2016
M. Conti, *Secure Wireless Sensor Networks*, Advances in Information Security 65,
DOI 10.1007/978-1-4939-3460-7_4

4.1 Introduction

Due to the operating nature of WSNs, they are often unattended, hence prone to different kinds of novel attacks. An adversary could capture nodes acquiring all the information stored therein—sensors are commonly assumed to be not tamper proof. Therefore, an adversary may clone captured sensors and deploy them in the network to launch a variety of malicious activities. This attack is referred to as the *clone attack* [41, 90, 247]. Since a clone has legitimate information (code and cryptographic material), it may participate in the network operations in the same way as a non-compromised node; hence cloned nodes can launch a variety of attacks. A few have been described in the literature [16, 217]. For instance, a clone could create a black hole, initiate a wormhole attack [121] with a collaborating adversary, or inject false data or aggregate data in such a way to bias the final result [235]. Further, clones can leak data.

The threat of a clone attack can be characterized by two main points:

- A clone is considered as a totally honest node from its neighbourhood. In fact, without global countermeasures, a honest node cannot be aware of the fact that it has a clone among its neighbours;
- to have a large amount of compromised nodes, the adversary does not need to compromise a high number of nodes. Indeed, once a single node has been captured and compromised, the main cost of the attack has been sustained. Making further clones of the same node can be considered cheap.

While centralized detection protocols have a single point of failure and high communication cost, local protocols do not detect cloned nodes that are distributed in different area of the network. In this chapter we look for a network self-healing mechanism, where nodes autonomously identify the presence of clones and exclude them from any further network activity. In particular, this mechanism is designed to iterate as a "routine" event: It is designed for continuous iteration without significantly affect the network performances, while achieving high clone detection rate.

In this chapter we analyze the desirable properties of distributed mechanisms for detection of node clone attack [52]. We also analyze the first protocol for distributed detection, proposed in [169], and show that this protocol is not completely satisfactory with respect to the above properties. Lastly, we propose a randomized, efficient, and distributed (RED) protocol for the detection of node clone attacks and we prove that our protocol meets all the above cited requirements. We further provide analytical results when RED and its competitor [169] face an adversary that selectively drops messages that could lead to clone detection. Finally, extensive simulations of RED show that it is highly efficient as for communications, memory, and computations required and shows improved attack detection probability (even when the adversary is allowed to selectively drop messages) when compared to other distributed protocols. We also propose some other distributed protocols for detection of clone attacks in [50, 54].

Organization

The remainder of this chapter is organized as follows: Next section reviews related work; Sect. 4.3 shows the threat model assumed in this chapter; Sect. 4.4 introduces the requirements a distributed protocol for the detection of the clone attack in wireless sensor networks should meet; in Sect. 4.5 we describe our randomized, efficient, and distributed solution; in Sect. 4.6 we show some experimental results on RED and compare them with the results obtained in [169]. These results confirm that RED matches the requirements in Sect. 4.4, that RED is more energy, memory, and computationally efficient, and that it detects node clone attacks with higher probability. In Sect. 4.7 we analyze how malicious nodes can affect the detection protocol performances. Finally, Sect. 4.8 presents some concluding remarks.

4.2 Related Work

One of the first solutions for the detection of clone attacks relies on a centralized base station [85]. In this solution, each node sends a list of its neighbours and their claimed locations (that is the geographical coordinates of each node) to a Base Station (BS). The same entry in two lists sent by nodes that are not "close" to each other will result in a clone detection. Then, the BS revokes the clones. This solution has several drawbacks, such as the presence of a single point of failure (the BS), and high communication cost due to the large number of messages. Further, nodes close to the BS will be required to route much more messages than other nodes, hence shortening their operational life.

Another centralized clone detection protocol has been proposed in [26]. This solution assumes that a random key pre-distribution security scheme is implemented in the sensor network. That is, each node is assigned a set of k symmetric keys, randomly selected from a larger pool of keys [85]. For the detection, each node constructs a counting Bloom filter from the keys it uses for communication. Then, each node sends its own filter to the BS. From all the reports, the BS counts the number of times each key is used in the network. The keys used too often (above a threshold) are considered cloned and a corresponding revocation procedure is raised.

Other solutions rely on local detection. For example, in [37, 79, 85, 159] a voting mechanism is used within a neighbourhood to agree on the legitimacy of a given node. However, this kind of a method, applied to the problem of clone attack detection, fails to detect clones that are not within the same neighbourhood. As described in [169], a naïve distributed solution for the detection of the node clone attack is Node-To-Network Broadcasting. In this solution each node floods the network with a message containing its location information and compares the received location information with that of its neighbours. If a neighbour s_w of node s_a receives a location claim that the same node s_a is in a position not coherent with the originally detected position of s_a, this will result in a clone detection. However, this method is very energy consuming since it requires N flooding per iteration, where N is the number of nodes in the WSN.

In the sybil attack [79, 159], a node claims multiple existing identities stolen from corrupted nodes. Note that both the sybil and the clone attacks are based on identity theft, however the two attacks are independent. The sybil attack can be efficiently addressed with mechanism based on RSSI [68] or with authentication based on the knowledge of a fixed key set [37, 46, 47, 75, 76].

Recent research threads cope with the more general problem of node compromise [53, 207, 241]. However, detecting node "misbehaviour" via an approach that is rooted on Intrusion Detection Systems theory [72] seems to require an higher overhead compared to clone detection. Indeed, in current solutions detecting a misbehaving node implies observing, storing, and processing a large amount of information. Some other recent studies on distributed detection of clone attack such as [199, 232, 245].

To the best of our knowledge the first not naïve, globally-aware, and distributed node-clone detection solution appeared in [169]. In particular, two distributed detection protocols leveraging emergent properties [98] have been proposed. The first one, the Randomized Multicast (RM), distributes node location information to randomly-selected nodes. The second one, the Line-Selected Multicast (LSM), uses the routing topology of the network to detect clones. In RM, when a node announces (locally broadcasts) its location, each of its neighbours sends (with probability p) a digitally signed copy of the location claim to a set of randomly selected nodes. Assuming that there is a cloned node, if every neighbour randomly selects $O(\sqrt{N})$ destinations, with a not negligible probability at least one node will receive a pair of not coherent location claims. We will call *witness* the node that detects the existence of a node in two different locations within the same protocol run. The RM protocol implies a high communication costs: Each neighbour has to send $O(\sqrt{N})$ messages. To solve this problem the authors propose the LSM Protocol.

The LSM Protocol is similar to RM but it introduces a remarkable improvement in terms of detection probability. In LSM, when a node announces its location, every neighbour first locally checks the signature of the claim and then, with probability p, forwards it to $g \geq 1$ randomly selected destination nodes. As an example, in Fig. 4.1 the node a announces its location and one of its neighbours, node b, forwards the

Fig. 4.1 Example of LSM Protocol iteration

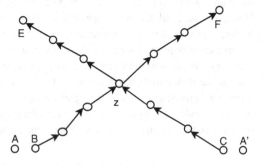

claim to node f. A location claim, when traveling from source to destination, has to pass through several intermediate nodes that form the so-called *claim message path*. Moreover, every node that routes this claim message has to check the signature, to store the message, and to check the coherence with the other location claims received within the same run of the detection protocol.

Node cloning is detected by the node (the witness, if present) on the intersection of two paths generated by two different node claims carrying the same *id* and coming from two different nodes. In the example shown in Fig. 4.1, node a' is a clone of node a (has the same *id* of node a). The claim of a' is forwarded by the node c to the node e. Node z will then results in the intersection of two paths carrying the claim of *id a* coming from different locations. Node z, the witness, will then trigger a revocation procedure.

In [244], the authors propose two different protocols with the aim of increasing the detection probability provided by LSM. The basic idea is to logically divide the network into cells and to consider all the nodes within a cell as possible witnesses. In the first proposed protocol, *Single Deterministic Cell*, each node *id* is associated to a single cell within the network. When the protocol runs, the neighbours of a node a probabilistically send the a's presence claim to the single pre-determined witness cell for a. Once the first node within that cell receives the claim message, it is flooded to all the other nodes within that cell. In the second proposal, *Parallel Multiple Probabilistic Cells*, the neighbours of a node a probabilistically send a's claim to a subset of the pre-defined witness cells for the id_a. The proposed solutions show an higher detection probability compared to LSM. However, the same predictable mechanism used to increase the detection probability can be exploited by the adversary for an attack—compromising the witnesses in order to go undetected. In fact, this predictability restricts the number of nodes (and their geographic areas) that can act as witnesses.

Another interesting distributed protocol for cloned node detection is the SET protocol [41]. SET leverages the knowledge of a random value broadcast by a BS to further perform a detection phase. In particular, the shared random value is first used to generate independent clusters and corresponding clusters heads. The specific clustering protocol used assures that the clusters are in fact Exclusive Subset Maximal Independent Set (ESMIS)—cluster heads are called Subset Leader (SLDRs). Further, within the same protocol iteration used to generate clusters and SLDRs, one or more trees are defined over the network graph. The nodes of the tree correspond to the SLDRs. Then, a bottom up aggregation protocol is run to aggregate the list of nodes belonging to the ESMIS. If a node *id* is present in two different independent subsets, than the node corresponding to that node *id* has been cloned.

The mechanism used by the protocol prevents a node to escape detection by claiming to be managed from a non existing SLDR—hence escaping the tree aggregation protocol. Note that defining such aggregating trees for each protocol iteration comes with a non-negligible cost in terms of messages. However, the main problem of this protocol is that the detection protocol itself is flawed—it can be maliciously exploited by the adversary to revoke honest nodes (that is, nodes that are not cloned). Indeed, a malicious node acting as a SLDR could declare in its ESMIS the presence of a honest node, say *a*, that eventually exists in some other part of the network (that is, belonging to a different ESMIS). This malicious behavior will lead the network to the "detection", and possibly to the revocation, of honest node *a*. Due to the possibility of this attack, in the following we do not consider SET as a benchmark for our protocol.

In [52] the desirable properties a clone detection protocol should meet has been pointed out. As shown in [52], the LSM Protocol [169] does not meet these properties. In particular, for LSM, some nodes have an higher probability to act as witnesses, then weakening the detection itself: The attacker can further take control of the node that will act as witness. Furthermore, the protocol's overhead is not evenly distributed among the network nodes. In [48] a randomized, efficient and distributed clone detection protocol has been proposed. The simulation results reported in [48] show that the proposed RED Protocol meets the desirable properties presented in [52].

In this chapter we review the contribution of [48] and further thoroughly investigate the feasibility of the RED Protocol. The analysis and the further set of simulations presented show that the RED Protocol can be actually implemented in sensor network. Also, it can be continuously iterated over the same network, as a self-healing mechanism, without significantly affecting the network performance (nodes energy and memory) and the detection protocol itself. Furthermore, we investigate the influence of an attacker intervening on message routing both for RED and its competing LSM Protocol.

4.3 The Threat Model

We define a simple yet powerful adversary: It can compromise a certain fixed amount of nodes and clone one or more into multiple copies (the *clones*). In general, to cope with this threat it could be possible to assume that nodes are tamper-proof. However, tamper proof hardware is expensive and energy demanding [3, 9]. Therefore, consistently with a large part of the literature, we will assume that the nodes do not have tamper proof components. The adversary goal is to prevent clones from being detected by the detection protocol used in the network. Hence, we assume that the adversary, to reach its goal, also tries to subvert the nodes that will possibly act as witnesses.

To formalize the adversary model, we introduce the following definition.

Definition 4.1 Assume that the adversary goal is to subvert the distributed detection protocol by compromising a possibly small subset T of the nodes. The adversary has already compromised a set of nodes $\mathscr{W}$, while $\mathscr{N}$ is the initial set of nodes in the network. For every node s, the *node appeal* $S(s)$ returns the probability that $s \in \mathscr{N} \setminus \mathscr{W}$ is a witness for the next run of the protocol.

We characterize the adversary through two different points of view: "where" and "how" it operates. As for "where", the adversary can be:

1. *localized*: The adversary chooses a convex sub-area of the network and compromises sensors from that area only.
2. *ubiquitous*: The adversary compromises sensors choosing from the whole network.

Intuitively, with the localized adversary we describe an adversary that needs some time to move from one point to another of the network area, while the ubiquitous adversary, during the same time interval, can capture nodes regardless of their position.

As for the sequence of node capture (that is, "how"), the adversary can be:

1. *oblivious*: At each step of the attack sequence, the next node to be tampered with is chosen randomly among the ones that are yet to be compromised;
2. *smart*: At each step of the attack sequence, the next node to tamper with is node s, where s maximizes $S(s)$, $s \in \mathcal{N} \setminus \mathcal{W}$.

Intuitively, the oblivious adversary does not take advantage of any information about the detection protocol implemented. Conversely, the smart adversary greedily chooses to compromise the node that maximizes its *appeal* in order to maximize the chance for its clones to go undetected.

4.4 Requirements for the Distributed Detection Protocol

In this section we present and justify the requirements a protocol for clone detection should meet.

4.4.1 Witness Distribution

A major issue in designing a protocol to detect clone attacks is the selection of witnesses. If the adversary knew the future witnesses before the detection protocol executes, the adversary could subvert these nodes for the attack to go undetected.

The adversary can in principle use any information on the network to foresee probability $S(s)$ for a generic node s. Here, we have identified the following two kinds of predictions:

- id-based prediction;
- location-based prediction.

We will say that a protocol for clone detection is *id oblivious* if the protocol does not provide any information on the *id* of the sensors that will be the witnesses of the clone attack during the next protocol run. Similarly, a protocol is *area oblivious* if probability $S(s)$, for every $s \in \mathcal{N}$, does not depend on the geographical position of node s in the network. Clearly, when a protocol is neither id-oblivious nor area-oblivious, then a smart adversary can have good chances of succeeding, since it is able to use this information to subvert the nodes that, most probably, will be the witnesses. Furthermore, when a protocol is not area oblivious, then even a localized oblivious adversary (that, at a first glance, seems to be the weakest) can enhance its chances of succeeding if it concentrates node compromising activities in an area with a high density of witnesses.

4.4.2 Overhead

Designing protocols for wireless sensor networks is a challenging task due to the resource constraints typical of these networks. Any protocol is required to generate little overhead. However, this requirement alone is not enough. Indeed, even if a protocol shows a reasonably small overhead on the average, it is still possible that a small subset of the nodes experiences a much higher overhead. This is bad—these nodes exhaust their batteries very quickly, with serious consequences on the network functionality. Moreover, the problem can be even more subtle when we consider memory. If a high memory overhead concentrates on a small number of nodes, then these nodes can overflow. During an overflow, the node could stop the protocol, or drop packets to free memory. It is very important to understand what kind of impact this scenario can have on the detection capability of the protocol itself.

We can summarize the above considerations with the general requirement that the overhead generated by the protocol should be small, that is sustainable by the network as a whole, and (almost) evenly distributed among the nodes. To make a real example, in LSM every node that relays a position claim must perform a signature verification and store the claim. As analyzed in [169], every line-segment includes $O(\sqrt{N})$ nodes and every node stores $O(\sqrt{N})$ location claims. Note that this memory requirement could be impractical in real networks with thousands of nodes.

Table 4.1 shows—first row—the asymptotic overhead for one protocol run (also referred to as *round* in the following) of LSM. The second row reports on the average overhead generated by one round of LSM for a network of 1,000 nodes with 31 neighbours per node (on the average). Finally, the third row shows the maximum overhead experienced by a node, that turns out to be much higher than the average. Detailed discussion on the generated overhead and compliance with the above described requirements are presented in Sect. 4.6.

Table 4.1 LSM overheads: $N = 1000$, $r = 0.1$, and $g = 1$

	Memory occupancy	Sent messages	Received messages	Signature check
Asymptotic	$O(g \cdot p \cdot d \cdot \sqrt{N})$	$O(g \cdot p \cdot d \cdot \sqrt{N})$	$O(g \cdot p \cdot d \cdot \sqrt{N})$	$O(g \cdot p \cdot d \cdot \sqrt{N})$
Average ($p = 0.1$)	20.33	22.08	49.84	21.08
Max ($p = 0.1$)	197	216	252	223
Average ($p = 0.05$)	9.98	10.98	38.60	10.17
Max ($p = 0.05$)	59	56	92	60

4.5 The RED Protocol

In this section we propose RED (Randomized, Efficient, and Distributed), a new protocol for the detection of clone attacks.

As in LSM, we assume that the nodes in the network are relatively stationary; that each node knows its own location (for instance, using a GPS or the protocols in [19, 31, 160]); and that all the nodes use a id-based public-key crypto system [42, 202]. We also assume that the network is loosely time synchronized. Observe that loose time synchronization can be achieved both in a centralized and in a distributed way [83, 84, 169].

RED executes routinely, at fixed intervals of time. Every run of the protocol consists of two steps.

In the first step a random value, *rand*, is shared among all the nodes. This random value can be broadcast with centralized mechanism (for example, from a satellite or a UAV [133], or other kinds of ground-based central stations), or with in-network distributed mechanisms. For instance, a non-subvertible, verifiable leader election mechanism [40, 71, 214] can be used to elect a leader among the nodes; this leader will later choose and broadcast the random value.

In the second step, each node digitally signs and locally broadcasts its claim—id and geographic location (Procedure BROADCAST_CLAIM shown in Algorithm 7). In the sequel of this chapter, without losing of generality and to ease exposition, we will rely on a centralized solution for the broadcast of the random value.

```
1 begin
2  │  claim ← ⟨id_a, is_claim, location(), time()⟩ ;
3  │  signed_claim ← ⟨claim, K_a^{priv}(claim)⟩ ;
4  │  a → NEIGHBOURS(): ⟨id_a, NEIGHBOURS(), signed_claim⟩ ;
5 end
```

Algorithm 7: BROADCAST_CLAIM.

When the neighbours receive the local broadcast, they execute the Procedure RECEIVE_MESSAGE (shown in Algorithm 8). Each of the neighbours sends (with probability p) the claim to a set of $g \geq 1$ pseudo-randomly selected network locations (rows 14–21 in Protocol 8). RED does not send the claim to a specific node id because this kind of a solution does not scale well: A claim sent to a node id that is no more present in the network would be lost; nodes deployed after the first network deployment could not be used as witnesses without updating all the nodes. However, RED can easily be adapted to work when a specific node is used as the message destination. Finally, in the following we consider the same geographic routing protocol used in [169] for a fair comparison. Though, RED is actually independent of the routing protocol used in the network.

```
 1 begin
 2 │ if IsCLAIM(M) then
 3 │ │   ⟨−, −, signed_claim⟩ ← M ;
 4 │ │   ⟨claim, signature⟩ ← signed_claim ;
 5 │ │   if BADSIGNATURE(claim, signature) then
 6 │ │   │   discard M ;
 7 │ │   else
 8 │ │   │   if INCOHERENTLOCATION(claim) then
 9 │ │   │   │   ⟨id_x, −, −, −⟩ ← claim ;
10 │ │   │   │   trigger revocation procedure for id_x ;
11 │ │   │   │   return ;
12 │ │   │   end
13 │ │   end
14 │ │   if with probability p then
15 │ │   │   ⟨claim, signature⟩ ← signed_claim ;
16 │ │   │   ⟨id_x, −, loc_x, time_x⟩ ← claim ;
17 │ │   │   locations ← PSEUDORAND(rand, id_x, g) ;
18 │ │   │   forall the l ∈ locations do
19 │ │   │   │   a → l: ⟨id_a, l, is_fwd_claim, signed_claim⟩ ;
20 │ │   │   end
21 │ │   end
22 │ else
23 │ │   if IsFWDCLAIM(M) then
24 │ │   │   ⟨−, −, −, signed_claim⟩ ← M ;
25 │ │   │   ⟨claim, signature⟩ ← signed_claim ;
26 │ │   │   if BADSIGNATURE(signed_claim) or REPLAYED(claim) then
27 │ │   │   │   discard M ;
28 │ │   │   else
29 │ │   │   │   ⟨id_x, −, loc_x, time_x⟩ ← claim ;
30 │ │   │   │   if DETECT_CLONE(memory, ⟨id_x, loc_x, time_x⟩) then
31 │ │   │   │   │   trigger revocation procedure for id_x ;
32 │ │   │   │   else
33 │ │   │   │   │   store fwd_claim in memory ;
34 │ │   │   │   end
35 │ │   │   end
36 │ │   end
37 │ end
38 end
```

Algorithm 8: RECEIVE_MESSAGE(M).

We assume that the routing delivers a message sent to a network location to the node closest to this location [31, 129]; that the routing protocol does not fail (as done in [169]); and that message forwarding is not affected by dropping or wormhole attacks (for these kinds of attacks a few solutions can be found in [67, 94, 128]). Later, in Sect. 4.7, we will see how the protocol performs when malicious nodes can drop packets. To test the protocol, we assume that the adversary has introduced two nodes with the same *id* in the network. Clearly, if the adversary introduces more clones of the same node, then the task of detecting the attack is only easier. Within this ideal framework, the probability that the clone attack is detected is equal to the probability that at least one neighbour of each clone sends the claim to the same witnesses. Considering d neighbours, the probability that from a neighbourhood a claim message is sent is $1 - (1 - p)^d$; therefore, the detection probability is $(1 - (1 - p)^d)^2$. For example, with $p = 0.1$ and $d = 35$ we have a detection probability of 0.95 in a single run of the protocol. Detection probability will be further discussed in a more realistic framework in Sect. 4.6.

The set of witnesses is selected using a pseudo-random function (line 17 of Protocol 8). This function takes in input the id of the node, that is the first argument of the claim message, the current $rand$ value, and the number g of locations that have to be generated. Using a pseudo-random function guarantees that, given a claim, the witnesses for this claim are unambiguously determined in the network, for a given protocol iteration. Time synchronization is used by the nodes to discern between different iterations.

Every node signs its claim message with its private key before broadcasting it (line 3 of Protocol 7). The nodes that forward the signed claim towards destination are not required to add any signature or to store any message. For every received claim, the potential witness node:

- Verifies the received signature (line 26);
- Checks for the freshness of message (line 26). This is important to prevent replay of old messages. This check is performed verifying the coherence between the time inserted in the message by the claiming node and the current time.

For every genuine message that passes the previous checks the witness node extracts the information (id and location). If this is the first claim carrying this id, then the node simply stores the message (line 33). If another claim from the same id has been received, the node checks if the new claim is coherent with the claim stored in memory for this id (line 30). If it is not, the witness triggers a revocation procedure for the id (line 31)—the two incoherent signed claims are the proof of cloning.

Here is an example of a run of the protocol. Assume that the adversary clones identity id_a and assigns this identity to nodes a' and a''. These two nodes are placed in two different network locations: l_1 and l_2 respectively. During a RED iteration, the nodes a' and a'' have to broadcast the same id, but different location claims (l_1 and l_2). Indeed, if $l_1 \sim l_2$, then either the neighbours of a' or the neighbours of a'' will raise an exception (line 10 of Protocol 8).

Let b' and b'' be neighbours of a' and a'', respectively. Using the pseudo-random function both b' and b'' will select the same set of witness nodes, containing at least a node w. In this way, w will receive two incoherent location claims for identity id_a— l_1 and l_2. This results in clone detection. Hence, w can start a revocation procedure for node id_a. Revocation can be performed by flooding the network with the two incoherent claims received by w. Remember that every claim message of a node is signed with the private key of the same node. Therefore, the two claims are a proof that id_a has been cloned.

The protocol shows one caveat: After the $rand$ value is shared, RED allows the adversary to know the witness set for any given id. However, note that the witnesses of a node could be anywhere in the network and that witnesses change at every protocol iteration in a unpredictable way. This means that the adversary, in order to prevent RED from detecting the clones, is required to be extremely fast and to capture all the witnesses of the clones within a window period that can be at most comprised between the disclosure of *rand* and the end of the protocol round. Considering realistic network sizes and the possible adversary speed, there are few chances for the adversary to perform this attack.

4.6 Simulations

In this section we show that RED meets the requirements described in Sect. 4.4: Area-obliviousness; id-obliviousness; low overhead; overhead balancing; and high clone attacks detection probability. We further compare RED with LSM and show that RED outperforms LSM in several ways.

In the following simulations we consider a unit square deployment area [20, 21, 76]. We fixed $N = 1,000$ nodes in the network and $r = 0.1$ communication radius. We also set $g = 1$ and $p = 0.1$ for both protocols. This means that the two protocols send the same number of location claims per node (on the average). Further, we assume that the nodes are distributed in the network area uniformly at random. We simulate the same geographic routing protocol used in [169]—the relay node is the neighbour closest to destination. The routing stops when no node is closer to destination than the current node: This node will be a witness. Note that this simple version of geographic routing, especially when used in networks that are sparse or deployed in an area that is not convex, has the problem of "dead ends"—places where the message cannot proceed because there is no node closer to destination, while the destination is still far. There are a few solutions to this issue [31, 81] that can be used in both protocols to guarantee that the claim reaches the node closest to destination.

The resources required by RED are shown in Table 4.2, for the same parameters used for LSM in Table 4.1. More details are given in the following sections.

Table 4.2 RED Overhead: $N = 1000$, $r = 0.1$, $g = 1$

	Memory occupancy	Sent messages	Received messages	Signature check
Asymptotic	$O(g \cdot p \cdot d \cdot)$	$O(g \cdot p \cdot d \cdot \sqrt{N})$	$O(g \cdot p \cdot d \cdot \sqrt{N})$	$O(g \cdot p \cdot d \cdot)$
Average (p = 0.1)	0.93	22.08	49.85	2.87
Max (p = 0.1)	15	220	250	48
Average (p = 0.05)	0.75	11.36	39.80	1.48
Max (p = 0.05)	6	67	98	11

4.6.1 Witness Distribution

Due to randomization, it is straightforward to verify that both LSM and RED are id-oblivious. In both protocols the ids of the witnesses are randomly selected among all the nodes in the network. To assess area-obliviousness, we study the witness distribution as follows: We select increasing sub-areas of the network, and for each sub-area we count the number of witnesses present in the area after a run of the detection protocol. Each sub-area from the center of the unit-square towards the external border provides an increment of 5% of the total area. Hence, 20 sub-areas are considered, as shown in Fig. 4.2.

In Fig. 4.3, we show an example of one iteration of LSM and RED. The black filled large circles indicate two clones of the same node, the gray filled small circles indicate nodes that route a claim from the clones, and finally the large not filled circle indicates the witness. This example suggests that LSM uses a higher number of routing nodes, compared to RED. Also, the witness nodes (large not filled circles) are located differently—near the center of the unit square for LSM while near the border for RED. In the following we will see through extensive simulations that this phenomenon is not episodical and we will analyze how it affects the performances of the protocol.

Figure 4.4 reports, for the two protocols, on the percentage of witnesses present in the incremental sub-areas. We simulate 10,000 different network deployments. For each deployment we randomly select two nodes, assign to them the same id, and execute a single LSM iteration and a single RED iteration. After each of these iterations we localize the witness nodes for the two different protocols. Finally, for

Fig. 4.2 Example of node deployment with 5% incremental sub-areas

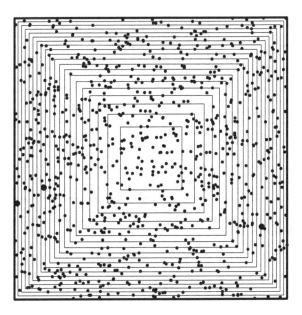

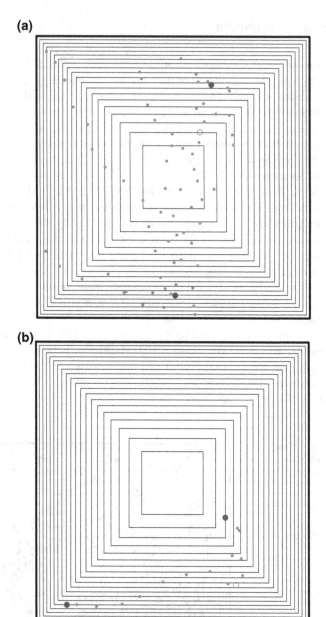

Fig. 4.3 Examples of protocols iteration. $N = 1000$, $r = 0.1$, $g = 1$, $p = 0.1$. **a** LSM Protocol.
b RED Protocol

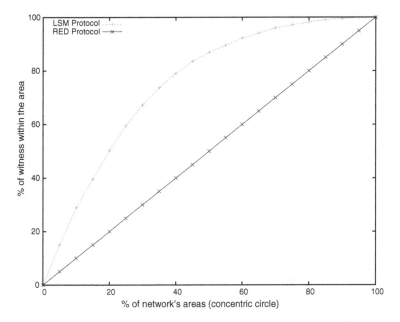

Fig. 4.4 Witness density. $N = 1000$, $r = 0.1$, $g = 1$, $p = 0.1$

each of the 20 incremental sub-areas we compute the percentage of witnesses with respect to the total number of witnesses. After collecting the outcome of 10,000 experiments, we plot the average. The x-axis of Fig. 4.4 indicates the percentage of the network area considered while the y-axis the corresponding percentage of all the witnesses in that area.

In Fig. 4.4 we can see that the central area, corresponding to 20 % of all the network area, collects more than 50 % of all the witnesses of LSM, while the most external area, corresponding to the 20 % of the network area, contains only 1.75 % of all the witnesses. Therefore, LSM is not area-oblivious, since $S(s_i) > S(s_j)$ for an s_i selected from the central area and an s_j selected from the most external area. Due to the pseudo-random choice of witness nodes in the RED Protocol, it is straightforward to prove that RED has a uniform witnesses distribution. In fact, Fig. 4.4 also shows how the behavior of RED corresponds to that of an ideal protocol: The witnesses are equally distributed in all the network areas. In other words, RED is area-oblivious.

4.6.2 Storage Overhead

Figure 4.5 reports the number of messages that the nodes are required to store for LSM and RED. For a fixed x-value of messages in memory, we show the percentage of the nodes that need to store that number of messages. The values were obtained averaging the result of 10,000 simulations.

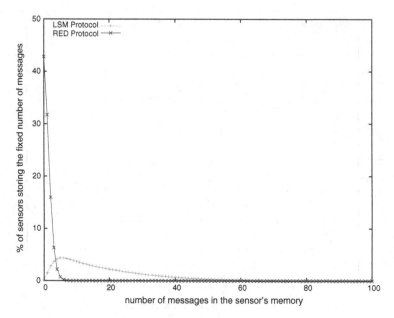

Fig. 4.5 Used memory for both RED and LSM. $N = 1000, r = 0.1, g = 1, p = 0.1$

Note that for LSM some nodes could require to store as many as 200 messages. Our experiments show that LSM requires some 1.9 % of the nodes to store more than 60 messages, some 7.6 % of nodes to store a number of messages between 40 and 59, and some 27.5 % of nodes to store a number of messages between 20 and 39. Just some 63 % of the nodes are required to store less than 20 messages.

As for RED, only a negligible percentage of nodes (0.001 %) require to store more than 10 messages. Moreover, some 0.3 % of the nodes need to store more than 5 messages and less than 10 % of nodes to store a number of messages between 3 and 5. It is interesting to note that 47.7 % of nodes need to store only one or two messages while 42.9 % of nodes do not require to store any message at all. Finally, observe that for LSM 0.2 % only of the nodes do not require to store any message. Figure 4.5 show memory requirements for the two protocols.

4.6.3 Energy Overhead

To assess the energy overhead of the two protocols we consider both communication and computation intensive operations (that is, public key cryptography: Signature generation and signature verification). In particular, we use the energy model proposed in [222]: A node battery of 324,000 mJ; 15.104 mJ for sending a packet and 7.168 mJ for receiving a packet (assuming packet length of 32 byte, 0.059 mJ for bit sending and 0.028 mJ for bit receiving); and 45.0 mJ for both signature generation and signature verification.

The operating life of a node depends on its battery. Different energy overheads for the two protocols will result in a different pattern of node exhaustion. Figure 4.6a shows this phenomenon. After 100 protocol run executed with the same network topology, some 20 % of the nodes are exhausted for LSM, while for RED all the nodes are alive. After 150 iterations, LSM shows 40 % of exhausted nodes, while RED only 5 %. Finally, after 200 run, LSM shows that half of the nodes of the network are exhausted (further, with such a number of exhausted nodes, the efficiency of LSM as for clone detection drops dramatically), while for RED this percentage is less than 15 % and the detection capabilities are still remarkable. It is also interesting to look at the different nodes exhaustion distribution in the network area. Figure 4.6b shows the distribution after 200 protocol iterations. The x-axis indicates the network sub-areas (as plotted in Fig. 4.2), numbered sequentially from the center (numbered 1) to the external one (numbered 20). The y-axis indicates the percentage of exhausted nodes in these areas. For both protocols most of the exhausted nodes are in the center. This phenomenon is known in the literature [135, 152] and it is due to the fact that most of the shortest paths generated by a uniform traffic traverse the center of the network. In the case of LSM, almost all the nodes in the center are exhausted (except a few isolated ones), and the overhead is transferred to the semi-central areas, leading to the shape in Fig. 4.6b.

Different distribution of node exhaustion also implies different clone attack detection probability, as shown in Fig. 4.7. This figure shows the detection probability (y-axis) at different protocol iterations (x-axis). In particular, we plotted the detection probability for the first 200 run. Plotted values were computed averaging the results obtained for 10,000 network deployments. Each single deployment was evaluated for both the LSM and the RED protocol. For all the considered iterations the RED Protocol shows a better detection probability compared to that of the LSM. From the 1st to the 50th iteration, LSM shows a probability detection of about 35 % while this probability is more than 80 % for the RED Protocol. It is interesting to note the tight relationship between the percentage of exhausted nodes (Fig. 4.6a) and the detection probability (Fig. 4.7). For LSM, nodes start exhausting after some 50 iterations; at the same iteration number, the detection probability starts decreasing. A similar behavior could be observed for RED as well. It is also possible to note that different slopes of the curves representing node exhaustion correspond to different slopes in curves representing detection probability.

Finally, we simulated the protocol behavior under a coordinated attack: The adversary clones a node into two copies and, in the same period of time, compromises a subset w of the other remaining nodes. In this setting, we assume that a compromised node forwards messages like a honest one: If not, this behavior could be detected, like in [67, 94, 128]. However, when a compromised node is a witness, we assume that it would not trigger any alarm, and the clones would go undetected for this specific protocol iteration. We investigated how the detection probability is affected under the above scenario, assuming that the adversary "smartly" compromises nodes from a so-called compromising area, which is a squared central area of the network—of increasing size.

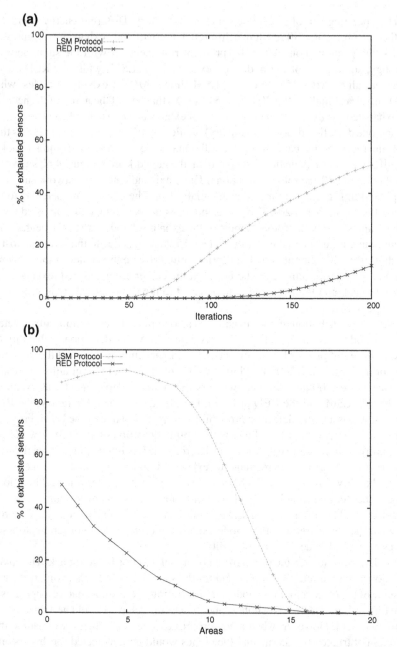

Fig. 4.6 Node exhaustion behavior: $N = 1000$, $r = 0.1$, $g = 1$, $p = 0.1$. **a** Exhausted node in different iterations. **b** Exhausted node distribution after 200 iteration

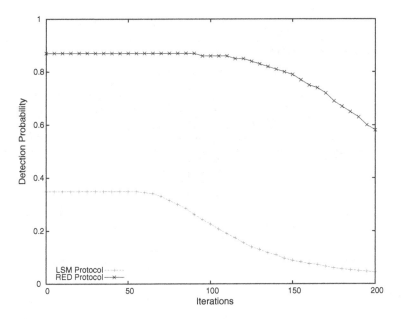

Fig. 4.7 Detection probability for both RED and LSM. $N = 1000, r = 0.1, g = 1, p = 0.1$

When no nodes are compromised, for the first 50 protocol run, the detection probability is 87 % for RED and 33.8 % for LSM, as shown in Fig. 4.7. When taking into account node compromising, the results of our simulations for LSM and RED are shown in Fig. 4.8a and b respectively. On the x-axis we indicate the number of compromised nodes while on the y-axis the percentage of the total network, starting from the inner area. We can notice that for LSM the detection probability is influenced by both the number of compromised nodes and the size of the compromising area (Fig. 4.8a). As for RED, the detection probability is influenced only by the number of compromised nodes (Fig. 4.8b). This is due to the following fact: As observed in Fig. 4.4, LSM shows an higher witness density in the most internal areas. For instance, capturing 150 nodes in the 20 % central area implies a reduction of detection probability of 25.4 % for LSM (from 33.8 to 25.2 %), while the performance of RED is reduced by 14 % only (from 87 to 74.8 %).

We can also note that, when the same number of nodes are compromised in all the network areas, the relative resilience of LSM is a little bit higher than RED. For example, with 150 compromised nodes all over the network area, LSM decreases its detection probability by 8.5 % only, while it is about 14 % for RED. This is due to the particular behavior of LSM: More that one node can witness a clone attack; compromising a witness node does not implies that a clone attack will go undetected for the LSM, while this can be true for RED. However, note that RED could be set to generate more than one witness. We decided to use the simplest version of RED to test RED under a constrained, very energy efficient scenario, since its detection performance is excellent even in this case.

(a)

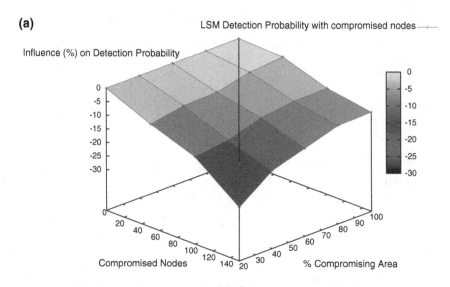

(b)

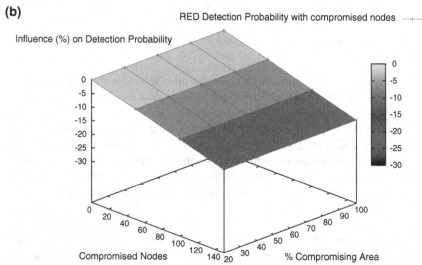

Fig. 4.8 Detection probability with compromised nodes. $N = 1000$, $r = 0.1$, $g = 1$, $p = 0.1$.
a LSM Protocol. **b** RED Protocol

4.7 Detection Probability with Malicious Nodes

In this section we investigate the clone detection probability during a sequence of
iterations. We assume that the adversary has cloned a node, that it is also controlling
a subset of w randomly selected other nodes, and that no mechanism for preventing
packet dropping is implemented, so that malicious nodes can stop claim forwarding.

We assume that a node (say a) is cloned and one of its clone (say a') is randomly deployed within the network area. We further assume no routing failure and that from each neighbourhood exactly one claim message is sent (we do not explicitly consider d, p and g values). Assume both claims are sent through path of length $\ell = c\sqrt{N}$ nodes (with constant network density, the average path length is $\Theta(\sqrt{N})$). The nodes on the two paths (the first one departing from the honest node, a, the second one from the clone, a') are those involved in the detection process by the two protocols.

In RED, if just one of these 2ℓ nodes in the two paths is malicious, detection can fail. In fact, note that the corrupted forwarding node can simply drop the received location claim. The probability that at least one malicious node is present in the two paths is:

$$1 - \frac{\binom{N-w}{2\ell}}{\binom{N}{2\ell}} \tag{4.1}$$

The probability that the attack is not detected using the RED Protocol, for a single protocol iteration, is exactly that of Eq. 4.1. To analyze a sequence of iterations, we assume that every iteration is probabilistically independent. Therefore, the probability that the attack is not detected after i RED Protocol iterations is:

$$\left(1 - \frac{\binom{N-w}{2\ell}}{\binom{N}{2\ell}}\right)^i. \tag{4.2}$$

The analysis is different for LSM. In fact, even if all the nodes are honest the attack is detected only with a given probability—the probability that two paths starting at a and a' intersect on a network node. Following the analysis proposed in [169], this probability is:

$$\frac{1}{3}\left(1 - \frac{35}{12\pi^2}\right). \tag{4.3}$$

However, note that the probability in Eq. 4.3 refers to geometric line intersection. Then, it is in fact an optimistic upper bound (also still assuming no failure in the routing). In fact, two intersecting paths (geometrically) do not necessarily have a node in common—an example of this case is shown in Fig. 4.9.

Despite this fact, in the following we optimistically consider Eq. 4.3 as the probability that the clone is detected when no malicious nodes are present.

Let U be the event that the attack is undetected for a single protocol iteration. For LSM we have to consider the following two disjoint events. Here, the idea is that malicious nodes can prevent clone detection only if they are in the path *before* the witness. Let us define:

- event E_h: All of the forwarding nodes before the (possibly present) witness are honest;
- event E_m: There is at least one malicious forwarding node before the (possibly present) witness.

Fig. 4.9 Example of
LSM-intersecting paths
without intersection node

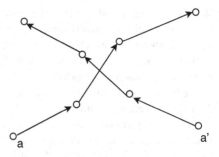

Note that E_h and E_m form a partition of the probability space, hence

$$Pr\,[U] = Pr\,[U|E_h]\,Pr\,[E_h] + Pr\,[U|E_m]\,Pr\,[E_m]\,. \tag{4.4}$$

$Pr\,[U|E_h]$ is the probability that the attack is undetected when there are no malicious nodes in the paths. According to [169], this is equal to

$$1 - \frac{1}{3}\left(1 - \frac{35}{12\pi^2}\right) = \frac{1}{3}\left(2 + \frac{35}{12\pi^2}\right). \tag{4.5}$$

We can assume that $Pr\,[U|E_m] = 1$, since the malicious node before the witness can discard the claim and stop the detection. $Pr\,[E_m] = 1 - Pr\,[E_h]$ is similar to Eq. 4.1, taking into account that the malicious nodes should appear before the witness. On the average, the witness is in the middle of the paths, therefore we can estimate this probability as follows:

$$Pr\,[E_m] = 1 - \frac{\binom{N-w}{\ell}}{\binom{N}{\ell}}.$$

Putting it altogether, we can compute $P\,(U)$ as follows:

$$Pr\,[U] = 1 + \frac{\binom{N-w}{\ell}}{\binom{N}{\ell}}\left(\frac{35}{36\pi^2} - \frac{1}{3}\right).$$

Therefore, the probability that the attack is not detected after i LSM Protocol iterations is:

$$\left[1 + \frac{\binom{N-w}{\ell}}{\binom{N}{\ell}}\left(\frac{35}{36\pi^2} - \frac{1}{3}\right)\right]^i. \tag{4.6}$$

Figure 4.10 shows the analytical results for RED and LSM on non-detection probability. Remind that, while the analysis for RED is essentially tight, the one for LSM

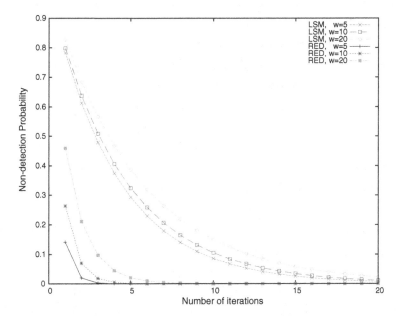

Fig. 4.10 Non-detection probability ($N = 1000$, $r = 0.5$)

is optimistic, since it depends on the assumption that paths that geometrically intersect have a node in common. This is not true, especially when the network is dense.
The actual detection rate depends on several factors like node density, for example.
Nonetheless, RED outperforms LSM even in the presence of malicious nodes that
can stop the protocol. Figures 4.11 and 4.12 shows the analytical results for several
values of c (c controls the length of the average random path in the network, being
$\ell = c\sqrt{N}$) of the non-detection probability. We considered subsequent protocol
iterations (x-axis). We plotted the result for $c = 0.1, 0.2, \ldots, 1$.

It is interesting to note how w and c influence the detection probability. Larger c
means longer paths and thus higher probability that one of the malicious nodes can
stop clone detection. Larger w means that the adversary can often thwart the protocols
and influence detection probability considerably, especially when c is large. In all
cases, it is clear that RED can converge to very high detection probability very quickly.
Note that RED is more influenced than LSM by path lengths, since a malicious node
can stop the protocol wherever it appears in the paths. However, experiments show
that, for a network of 1000 nodes and communication range 0.1 in a network area
of side 1, c is about 0.35. Therefore, we can conclude that RED has better detection
probability and converges faster than LSM for all practical values of the network
parameters.

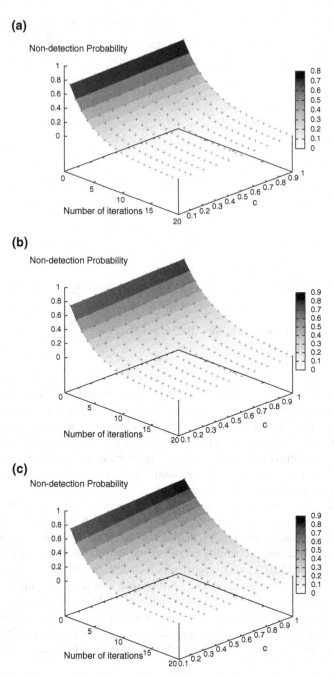

Fig. 4.11 Non-detection probability for LSM. **a** $w = 5$, **b** $w = 10$, **c** $w = 20$

(a)

Non-detection Probability

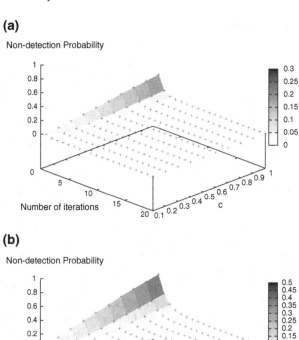

(b)

Non-detection Probability

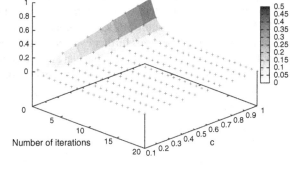

(c)

Non-detection Probability

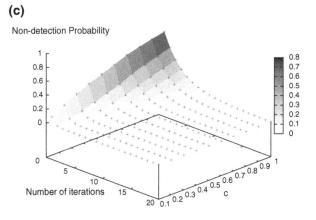

Fig. 4.12 Non-detection probability for RED. **a** $w = 5$, **b** $w = 10$, **c** $w = 20$

4.8 Concluding Remarks

In this chapter we presented and justified a few basic requirements an ideal protocol for distributed detection of cloned nodes should have. In particular, we have introduced the preliminary notion of *id-obliviousness* and *area-obliviousness* that convey a measure of the quality of the clone attack detection protocol; that is, its resilience to a smart adversary. Moreover, we have indicated that the overhead of such a protocol should be not only small, but also evenly distributed among the nodes, both in computation and memory. Further, we have introduced new adversary threat models. However, a major contribution of this chapter is the proposal of a randomized, efficient, and distributed protocol (RED) to detect node clone attacks. We analytically compared the RED Protocol with the state of the art solution (LSM) and proved that the overhead introduced by RED is low and almost evenly balanced among the nodes; RED is both id-oblivious and area-oblivious; furthermore, RED efficiency and effectiveness outperforms LSM. Extensive simulations confirm these results. Finally, also when coping with an adversary that uses compromised nodes—affecting routing—to stay undetected, RED is more resilient in its detection capabilities than LSM. We leave as a future work the implementation of both RED and LSM on sensor networking platform for further evaluation and comparisons.

This chapter addressed a general type of attack that can affect different WSN services (communication services, voting mechanisms, data collection, data processing, etc.). The next chapter deals with an attacker that wants to exploit a specific WSN service, the data collection.

Chapter 5
Secure Data Aggregation

Chapter 4 considered a general WSN security problem. In this chapter we want to focus on a specific WSN service: The data collection. If each single node sends its own data to the collecting point in an independent way this will result in a huge waste of energy. To meet the severe energy constraints in WSNs, some researchers have proposed to use the in-network data aggregation technique (i.e., combining partial results at intermediate nodes during message routing), which significantly reduces the communication overhead. Some researchers also proposed algorithms to securely compute a few aggregates, such as Sum (the sum of the sensed values), Count (number of nodes) and Average. However, to the best of our knowledge, there is no prior work which securely computes the Median, although the Median is considered to be an important aggregate. The contribution of this chapter is twofold. We first propose a protocol to compute an approximate Median and verify if it has been falsified by an adversary. Then, we design an attack-resilient algorithm to compute the Median even in the presence of a few compromised nodes. We evaluate the performance and cost of our approach via both analysis and simulation. Our results show that our approach is scalable and efficient.

5.1 Introduction

The simplest way to collect the sensed data in a WSN is to let each sensor node deliver its reading to the base station (BS). This approach, however, is wasteful since it results in excessive communication. A typical sensor node is severely constrained in communication bandwidth and energy reserve. Hence, sensor network designers have advocated alternative approaches for data collection.

An *in-network* aggregation algorithm combines partial results at intermediate nodes during message routing, which significantly reduces the amount of communication and hence the energy consumed. A typical data acquisition system [91,

© Springer Science+Business Media New York 2016
M. Conti, *Secure Wireless Sensor Networks*, Advances in Information Security 65,
DOI 10.1007/978-1-4939-3460-7_5

148] constructs a spanning tree rooted at the BS and then performs in-network aggregation along the tree. Partial results propagate level by level up the tree, with each node awaiting messages from all of its children before sending a new partial result to its parent. Researchers [91, 148] have designed several energy-efficient algorithms to compute aggregates such as Count, Sum, Average, etc. However, an in-network aggregation algorithm cannot cheaply compute the exact Median, where the worst case communication overhead per node is $\Omega(N)$, where N is the number of nodes in the network [148]. As a result, researchers have advocated computation of an approximate Median. In-network aggregation algorithms to compute an approximate Median are proposed in [101, 204].

Unfortunately, none of the above algorithms include any provisions for security, and hence, they cannot be used in security-sensitive applications. Given the lack of tamper-resistance and the unattended nature of many networks, we must consider the possibility that a few sensor nodes in the network might become compromised.

A compromised node in the aggregation hierarchy may attempt to change the aggregate value computed at the BS by relaying a false sub-aggregate value to its parent. This attack can be launched on most of the in-network aggregation algorithms. For example, in Greenwald et al.'s approximate Median computation algorithm [101], a compromised node in the aggregation hierarchy can corrupt the quantile summary to make the BS accept a false Median which might contain a large amount of error.

A technique to compute and verify Sum and Count aggregates has been proposed by Chan et al. [38]. Their scheme [38] can also verify if a given value is the true Median, but they have not proposed any solution to compute that value in the first place. To the best of our knowledge, there is no prior work which securely computes the Median using an in-network algorithm.

One might suggest an approach which runs Greenwald et al.'s algorithm [101] to compute an approximate Median and then employs Chan et al.'s verification protocol [38] to verify if the computed value is indeed a valid estimate. We refer this approach as GC in the rest of the chapter. The communication cost per node in this approach is $O(\frac{log^2 N}{\epsilon})$, where ϵ is the approximation error bound.

In this chapter, we propose an alternative approach to compute and verify an approximate Median, which proves to be more efficient compared to the GC approach. Our approach is based on sampling—an uniform sample of sensed values is collected from the network to make a preliminary estimate of the Median, which is verified and refined later. The communication cost of our basic algorithm is $O(\frac{1}{\epsilon}\Delta \log N)$, where ϵ is the error bound and Δ is the maximum degree of the aggregation tree used by the algorithm.

Like the GC approach, our basic algorithm guarantees that an attacker cannot cause the BS to accept a Median estimate which contains an error more than the user-specified bound, ϵ. However, neither of the above approaches can guarantee the successful computation of the Median in the presence of an attacker. To address this problem, we extend the basic approach so that we can compute the Median even in the presence of a few compromised nodes. The analysis and simulation results

Table 5.1 Median computation protocols: Comparing the performance and the security features

	Node congestion	Latency (epochs)	Verification	Attack-resilient computation
Greenwald et al.'s protocol [101]	$O((log^2 N)/\epsilon)$	2	No	No
GC approach (Sect. 5.4.1)	$O((log^2 N)/\epsilon)$	6	**Yes**	No
Our basic protocol (Sect. 5.4.3)	$O((1/\epsilon)\Delta log N)$	6 w.h.p.	**Yes**	No
Our extended protocol (Sect. 5.6)	$O((1/\epsilon)\Delta log N)$	6 w.h.p.	**Yes**	**Yes**

show that our algorithms are effective and efficient. Further, our algorithms can be extended to compute other quantiles.

Table 5.1 compares our approach with other solutions on the basis of a few performance and security metrics. We report *node congestion* as a metric for communication complexity, which represents the worst case overhead on a single node. We measure the latency of the protocols in *epochs*. Similarly to the prior work [148], an epoch represents the amount of time a message takes to traverse the distance between the BS and the farthest node on the aggregation hierarchy. We observe that the latency of our protocol might increase in extreme cases; here we report the latency which our protocol incurs in most cases (i.e., with high probability (w.h.p.)). To measure the security of the protocols, we consider the following properties. We say that a protocol has *verification* property if the protocol enables the BS to verify whether the computed Median is false or not. Observe that this property does not guarantee the computation of the Median in the presence of an attack. Finally, an attack-resilient protocol is so if it guarantees the computation of the Median in the presence of a few malicious nodes. We further investigated data aggregation security considering the attacker's impact in our recent works [193, 194].

Organization

The rest of the chapter is organized as follows. In Sect. 5.2, we review the related work present in the literature. Section 5.3 describes the problem and the assumptions taken in this chapter. In Sect. 5.4, we present our basic protocol, whose security and performance analysis is given in Sect. 5.5. Section 5.6 describes our attack-resilient protocol. We present our simulation results in Sect. 5.7, and finally, we conclude the chapter in Sect. 5.8.

5.2 Related Work

Several researchers [91, 148] have proposed in-network aggregation algorithms
which fuse the sensed information en route to the BS to reduce the communication
overhead. In particular, these algorithms are designed to compute *algebraic aggregates*, such as Sum, Count, and Average. However, Madden et al. [148] showed
that in-network aggregation does not save any communication overhead in case of
computing *holistic* aggregates, such as the Median.

To limit the communication complexity, researchers have advocated computing
an approximate estimate instead of the exact Median [101, 204]. In particular, Greenwald et al. [101] proposed a *quantile summary* computation algorithm that exploits
a concept of *delayed aggregation* so that no summary contains error more than ϵ
bound. Also, Srivastava et al. [204] presented another data summarization technique
called *quantile digest* to compute an approximate Median, where the main idea is to
compute an equi-depth histogram through in-network aggregation. There also exists
a body of data stream algorithms in the literature which computes approximate quantiles [61, 100, 150]. In fact, Greenwald et al.'s algorithm [101] is an extension of
[100].

Our Median computation algorithm has a sampling phase and a histogram computation phase. Sampling techniques have been previously employed for data reduction
in databases [15, 179]; in particular, [15] uses a sample of a large database to obtain
an approximate answer. Another work, from Munro and Paterson [156], analyzed
the lower bound on storage space and number of passes of a Median computation
algorithm. Jain et al. [126] proposed a centralized algorithm to compute quantiles and
histograms with limited storage space. Patt-Shamir [170] designed an approximate
Median computation algorithm using the synopsis diffusion framework [43, 158],
which uses a multipath routing algorithm to enhance robustness against communication loss. We note that none of the above algorithms were designed with security
in mind, and an attacker can inject an arbitrary amount of error in the final estimate.

Recently, researchers have considered security issues in aggregation algorithms.
Boubiche et al. proposed SDAW [24], an energy efficient secure data aggregation
approach based on watermarking. In [186], Rezvani et al. focused on iterative filtering algorithms for data aggregation and proposed a collusion-resistant approach
by providing an initial approximation of the trustworthiness of the nodes. Wagner [220] addressed the problem of resilient data aggregation in the presence of
malicious nodes and provided guidelines for selecting aggregation functions in a
sensor network. Yang et al. [235] proposed SDAP, a secure hop-by-hop data aggregation protocol using a tree-based topology to compute the Average in the presence
of a few compromised nodes. SDAP divides the network into multiple groups and
employs an outlier detection algorithm to detect the corrupted groups. In our extended
approach, we also use a grouping technique but without any outlier detection algorithm that would otherwise require the assumption that groups have similar data distribution. Another approach for the securely computing Count and Sum, proposed by
Roy et al. [195], is designed for the synopsis diffusion framework [43, 158].

Chan et al. [38] designed a verification algorithm by which the BS could detect if the computed aggregate was falsified. However, the authors did not propose any algorithm to compute the Median. Their verification algorithm is based on a novel method of distributing the verification responsibility onto the individual sensor nodes. An improvement on the communication complexity of the above algorithm has been proposed by Frikken et al. [89].

In this chapter we often refer to Greenwald et al.'s Approximate Median Algorithm [101] and to the Chan et al.'s Verification Algorithm [38]. To help the reader, we briefly present these algorithms in the following part of this section.

5.2.1 Greenwald et al.'s Approximate Median Algorithm

This algorithm [101] is based on a summarization technique which represents a set of sensor readings by a *quantile summary*. From a ϵ-approximate quantile summary, we can derive an arbitrary quantile of the data set satisfying ϵ-approximation error bound. In particular, an ϵ-approximate quantile summary for a data set A is an ordered set $Q = \{\alpha_1, \alpha_2, \ldots, \alpha_l\}$ such that (i) $\alpha_1 \leq \alpha_2 \ldots \leq \alpha_l$ and $\alpha_i \in A$ for $1 \leq i \leq l$, and (ii) $rank\,(\alpha_i + 1) - rank\,(\alpha_i) < 2 \cdot \epsilon \cdot |A|$.

Also, given two quantile summaries, Q_1 and Q_2, which represent two disjoint sets of sensed values, A_1 and A_2, respectively, we can aggregate them into a single quantile summary Q which represents all the values in $A = A_1 \cup A_2$. To aggregate two quantile summaries, we need two operations: *combine operation* and *prune operation*. The output of the combine operation from the quantile summaries Q_1 and Q_2 is a sorted list, Q', which is the union of Q_1 and Q_2. As a result, the size of Q' is the sum of the sizes of the original summaries Q_1 and Q_2. To keep the size of the quantile summary within limits, we apply the prune operation on Q' to determine a quantile summary Q of a constant size, say z. The prune operation introduces an additional error to that contained in the original summary. In particular, if ϵ' is the error in Q', then the error in Q will be $\epsilon' + \frac{1}{2z}$.

The aggregation of individual quantile summaries is performed over a tree structure with the BS as the root, which is formed in the query broadcast phase. A leaf node sends its quantile summary, which is simply its sensed value, to its parent. Each non-leaf node X first aggregates the quantile summaries it receives from its child nodes using the *combine* operation, and finally X applies one *prune* operation to keep the size of the summary constant. Due to the error introduced by the *prune* operation, the algorithm uses a concept of delayed aggregation, where the number of prune operations is kept within limit to satisfy the error bound ϵ in the final quantile summary. The authors design the protocol in such a way that a single sensed value experiences at most $\log N$ number of prune operations on its way to the BS. If we set the quantile size $z = \frac{\log N}{\epsilon}$, then the final error is bound to be ϵ and the worst case node congestion is $O(\frac{\log^2 N}{\epsilon})$.

5.2.2 Chan et al.'s Verification Algorithm

This scheme [38] is designed to compute and verify the Sum aggregate. The main idea behind this scheme is to move the verification responsibility from the BS to individual nodes that participated in the aggregation. Each node verifies if its own value is accounted for in the final aggregate. The algorithm consists of four operations, each of which takes one epoch to complete: (i) query dissemination, (ii) aggregation-commit, (iii) commitment-dissemination, and (iv) result-checking.

In the first epoch, the BS broadcasts an aggregation request. As the query message propagates through the network, an aggregation tree with the BS at the root is formed like in TAG algorithm [148].

During the aggregation-commit epoch, while the Sum is computed over an aggregation tree, nodes also construct a commitment structure similar to a Merkle hash tree [153] to enable the verification in the next phase. While a leaf node's message to its parent node contains its sensed value, each internal node sends the sub-aggregate it computed using the values received from its child nodes. In addition, each internal node, X, creates a commitment (a hash value) of the messages received from its child nodes. Both the sub-aggregate and the commitment are then passed to X's parent, which acts as a summary of X's sub-tree. The fields in X's message are $< \beta, v, \bar{v}, h >$, where β is the number of nodes in X's sub-tree, v is the local sum, $\bar{v}$ is the complement of the local sum (considering an upper bound v_{bound} for a sensed value), and h is an authentication field. In particular, a leaf node X sets the fields in its message as follows: $\beta = 1, v = v_X, \bar{v} = v_{bound} - v_X$, and $h = X$. If an internal node X receives messages $u_1, u_2, \ldots, u_t$ from its t child nodes, where $u_i = < \beta_i, v_i, \bar{v}_i, h_i >$, then X's message, $< \beta, v, \bar{v}, h >$, is generated as follows: $\beta = \sum \beta_i + 1, v = \sum v_i + v_X$, $v = \sum \bar{v}_i + (v_{bound} - v_X)$, and $h = H[\beta || v || \bar{v} || u_1 || u_2 || \ldots || u_t]$, where H is a hash function. Once the BS receives the final commitment, it verifies the coherence of the final $v, \bar{v}$ with N number of nodes in the network, and the upper bound of sensed value, v_{bound}. In particular, the BS performs the following sanity check: $v + \bar{v} = v_{bound} \times N$. If this check succeeds, the base station initiates the next phase.

In the commitment-dissemination epoch, the final commitment C is broadcast by the BS to the network. This message is authenticated using the $\mu Tesla$ protocol [172]. The aim of the commitment dissemination phase is to let each single node know that its own value has been considered in the final aggregate. To do so, each node X should receive all of the *off-path values* up to the root node relative to X's position on the commitment tree. These values, together with the X's local commitment, allows X to compute a final commitment C'. Finally, node X checks if $C' = C$. If the check succeeds, it means that X's local value, v_X, has been included in the final Sum received by the BS.

In the last epoch, each node X that succeeded in the previous check sends an authentication code (MAC) up the aggregation tree toward the BS. These MACs are aggregated along the way with the XOR function to reduce the communication overhead. When the BS receives the XOR of all of the MACs, it can verify if all nodes confirmed that their values have been considered in the final aggregate.

The main cost of this protocol is due to the dissemination of the off-path values to individual nodes. The authors observed that this overhead is minimized if the commitment structure is balanced. They proposed to decouple the commitment structure from the physical aggregation tree, which enables the building of a balanced commitment forest as an overlay on an unbalanced aggregation tree. That results in the worst case node congestion in the protocol being $O(\Delta \log^2 N)$. To further reduce this overhead, Frikken et al. [89] modified the commitment structure, which results in a total cost of $O(\Delta \log N)$.

Finally, the authors show how the Sum computation protocol can be extended to compute the cardinality of a subset of nodes (Count) in the network. In particular, to count the elements in a given subset, we require each node to contribute 1 to the Sum aggregate if it belongs to the subset and to contribute 0 otherwise.

5.3 Assumptions and Problem Description

The goal of this chapter is to securely compute an approximate Median of the sensor readings in a network where a few nodes might be compromised. Given a specified error bound, we return an approximate Median which is sufficiently close to the exact Median. This section describes our system model and design goals.

Network Assumptions. We assume a general multihop network with a set of N sensor nodes and a single BS. The BS knows the IDs of the sensor nodes present in the network. The network user controls the BS, initiates the query and specifies the error bound ϵ. In the rest of the chapter, we consider the user and the BS as a single entity. We also consider that sensor nodes are similar to the current generation of sensor nodes (e.g., Berkeley MICA2 motes [119]) in their computational and communication capabilities and power resources, while the BS is a laptop-class device supplied with long-lasting power.

We assume that the in-network aggregation is performed over an *aggregation tree* which is constructed during the query broadcast, similarly as in the TAG algorithm [148]. However, our approach does not rely on a specific tree construction algorithm. The approximation error ϵ in the estimated Median $\hat{m}$ is determined by how many position $\hat{m}$ is away from the exact Median m in the sorted list of all the sensed values. For ease of exposition, without loss of generality we assume that all the sensed values are distinct. Note that we could relax this assumption by defining an order on the nodes' ID that have same sensed value. Also, for the ease of exposition, we assume that there is an odd number of sensed values in total so that the Median is one element of the population.

Security Model. We assume that the BS cannot be compromised. The BS uses a protocol such as $\mu Tesla$ [172] to authenticate broadcast messages. We also assume that each node X shares a pairwise key, K_X with the BS, which is used to authenticate the messages it sends to BS.

In this chapter, we do not address outsider attacks—we can easily prevent unauthorized nodes from launching attacks by augmenting the aggregation framework with authentication and encryption protocols [172, 248].

We consider that the adversary can compromise a few sensor nodes (i.e., insiders) without being detected. If a node is compromised, all the information it holds will also be compromised. We use a Byzantine fault model, where the adversary can inject malicious messages into the network through the compromised nodes. We observe that a compromised node might launch multiple potential attacks against a tree-based aggregation protocol, such as corrupting the underlying routing protocol, selective dropping, or a Denial of Service attack to prevent other nodes from receiving the messages from the BS. However, in this chapter we address only false data injection attacks where the goal of the attacker is to cause the BS to accept a false aggregate. To achieve this goal in an in-network Median computation algorithm (e.g., [101]), a compromised node X could either attempt to falsify its own sensed value, v_X, or the sub-aggregate X is supposed to forward to its parent. We note that as we are computing Median, by falsifying the local value a compromised node can only deviate the final estimate by one position, i.e., the impact of the *falsified local value attack* is very limited. Moreover, it is impossible to detect the falsified local value attack without domain knowledge about what is an anomalous sensor reading. On the other hand, the *falsified sub-aggregate attack*, in which a node X does not correctly aggregate the values received from X's child nodes, poses a large threat to an in-network Median computation algorithm; a compromised node X forwards to its parent a corrupted aggregate which falsely summarizes X's descendants' sensed values. We observe that by launching this attack, a single compromised node, which is placed near the root on the aggregation hierarchy, can deviate the final estimate of the Median by a large amount (e.g., in [101]).

Problem Description. We aim to compute an approximate Median against the *falsified sub-aggregate attack*. In particular, our goal is to design the following two algorithms.

- Median computation and verification algorithm: This algorithm either outputs a valid approximate Median or it detects the presence of an attack. A value, $\hat{m}$, is considered to be a valid approximate Median if it is close to the exact Median, m, within the bound specified by the user. In particular, if the user-specified relative error bound is ϵ, the BS accepts an estimate $\hat{m}$ which satisfies the following constraint:

$$|rank\,(\hat{m}) - \frac{N+1}{2}| \leq \epsilon N \tag{5.1}$$

  where $rank\,(x)$ denotes the position of the value x in the sorted list of all the sensed values (the population elements), and N is the size of the population.
- Attack-resilient Median computation algorithm: If the above verification fails, our further aim is to compute an approximate Median in the presence of the attack.

We finally note that by launching a *falsified local value attack*, w compromised nodes can deviate $rank\,(\hat{m})$ in constraint (1) above by w positions, which makes the

Table 5.2 Secure Median computation: Notations

Symbol	Meaning
N	Total number of nodes (or total number of sensed values)
S	Sample size
E_i	Value of ith item in the sorted sample
K_X	Symmetric key shared between node X and the BS
ϵ	Error bound for the approximate Median
q_i	Bucket boundary in histogram
$B_i \equiv [q_i, q_{i+1}]$	ith bucket of the histogram
c_i	Count of ith bucket
v_X	Sensed value of node X
$MAC(K_X, M)$	Message authentication code of message M computed using key K_X
V_X	$=(X, v_X, MAC(K_X, v_X))$
$X \rightarrow Y$	X sends a message to Y
$X \rightarrow *$	X broadcasts a message
$X \Longrightarrow Y$	X sends a message to Y via multiple paths
$a_1 \parallel a_2$	Concatenation of string a_1 and a_2
Δ	The maximum degree of the aggregation tree
g	Number of groups in the attack-resilient algorithms
w	Number of compromised nodes

error bound of the final estimate of the Median to be $(\epsilon + w/N)$. However, given an upper bound on w, the user could adjust his input ϵ to finally meet the required bound.

Notation. A list of notations used in this chapter is given in Table 5.2.

5.4 Computing and Verifying an Approximate Median

The key elements of our approach are to compute a histogram of the sensor readings and then derive an approximate Median from the histogram. We collect a sample of sensed values from the network which is used to construct the histogram bucket boundaries. Before we present our scheme, we first discuss an approach to securely compute an approximation Median whose performance will be later compared with that of our scheme. Then, we present a histogram verification algorithm and finally describe our basic scheme.

5.4.1 GC Approach

One can suggest a scheme to securely compute an approximate Median using Greenwald et al.'s Median computation algorithm [101] in conjunction with Chan et al.'s verification algorithm [38]. A brief description of these algorithms can also be found in Sects. 5.2.1 and 5.2.2. In the first phase of GC approach, given the approximation error bound ϵ, we can run Greenwald et al.'s algorithm to compute a quantile summary. From the quantile summary we can derive an approximate Median $\hat{m}$ which is supposed to satisfy ϵ error bound. In the next phase, we can verify the actual error present in the estimate, $\hat{m}$, which might have been falsified by an attacker in the previous phase. To verify the error, Chan et al.'s verification algorithm can be applied to count the number of nodes in the network whose value is no more than $\hat{m}$.

The communication cost per node in this approach comes from the original protocols: That is $O(\frac{log^2 N}{\epsilon})$ for Greenwald et al.'s Median computation algorithm and $O(\Delta \log N)$ for Chan et al.'s verification scheme (considering Frikken et al.'s improvement [89]), where N is the number of nodes in the network, ϵ is the approximation error bound and Δ is the number of neighbours of a node.

5.4.2 A Histogram Verification Algorithm

We now present an algorithm for computing and verifying a histogram of sensed values, which is adapted from Chan et al.'s scheme [38] to compute and verify Sum aggregate.

Formally speaking, a histogram is a list of ordered values, $\{q_0, q_1, \ldots, q_i, \ldots\}$, where each pair of consecutive values (q_i, q_{i+1}) is associated with a count c_i which represents the number of population elements, v_j, such that $q_i < v_j \leq q_{i+1}$. We refer such an interval, (q_i, q_{i+1}) as bucket B_i with boundaries q_i and q_{i+1}.

As noted in [38], the Sum scheme can be adapted to count the cardinality of a subset of nodes. Here, we apply Sum aggregate to count how many sensor readings belong to each histogram bucket. To do so, we require each node X to contribute 1 to the count of its corresponding bucket (the bucket X's sensed value, v_X, lies within) in the histogram while we compute the total count for each bucket. Like Chan et el.'s scheme, the histogram verification scheme takes four epochs to complete: Query dissemination, aggregation-commit, commitment-dissemination, and result-checking.

After an aggregation tree is constructed in the query broadcast epoch, each node X's message in the aggregation-commit epoch looks like $< \beta, c_1, c_2, \ldots, c_b, h >$, where β is the number of nodes in X's subtree, b is the number of buckets in the histogram, each c_i represents the count for the bucket B_i, i.e., $\beta = \sum_i c_i$, and h is an authentication field. Note that for each bucket count c_j all of the other bucket counts together act as a complement, i.e., $c_j + \sum_{i \neq j} c_i = \beta$. A leaf node X whose sensed value, v_X, lies within the bucket B_j sets the fields in its message as follows: $\beta = 1, c_j = 1, c_i = 0$ for all $i \neq j$, and $h = X$. If an internal node X whose value v_X lies within the bucket B_j receives messages $u_1, u_2, \ldots, u_t$ from its t child nodes, where $u_k = < \beta_k, c_1^k, c_2^k, \ldots, c_b^k, h_k >$, then X's message $< \beta, c_1, c_2, \ldots, c_b, h >$

Fig. 5.1 The aggregation-commit phase in histogram verification: In this example, v_X lies in bucket B_2, v_Y lies in bucket B_1, and v_Z lies in the last bucket B_b

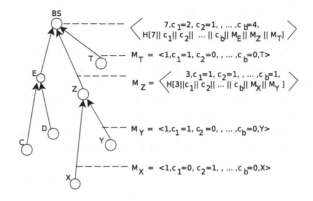

is generated as follows: $\beta = \sum \beta_k + 1$, $c_1 = \sum c_1^k$, $c_2 = \sum c_2^k, \ldots, c_j = \sum c_j^k + 1, \ldots, c_b = \sum c_b^k$, and $h = H[\beta||c_1||c_2||\ldots||c_b||u_1||u_2||\ldots||u_t]$, where H is a hash function. The above messages along the aggregation hierarchy logically build a commitment tree which enables the authentication operation in the next phase. Once the base station receives the final commitment, it verifies the coherence of the final counts, $c_1, c_2, \ldots, c_b$, with the number of nodes in the network, N. In particular, the BS performs the following sanity check: $\sum c_i = N$. A simplified version of the aggregation-commit phase is illustrated in Fig. 5.1 with an example of a small network.

Both the commitment-dissemination epoch and the result-checking epoch are straightforward extensions of those in Chan et al.'s Sum scheme. During commitment-dissemination epoch, the final commitment is broadcast by the BS to the network. In addition, each node X receives from its parent node all of the *off-path values* up to the root relative to X's position on the commitment tree. The aim of the com-mitment dissemination phase is to let each single node know that its own value has been considered in the final histogram. The message containing the *off-path values* received by a node is bigger compared to that in the Sum scheme because each off-path value contains b counts when a histogram with b buckets is computed. In the result-checking epoch, the BS receives a compressed authentication code from all of the nodes which enables to verify if each node confirmed that its value has been considered in the final histogram.

As in Chan et al.'s Sum scheme, the main cost of this protocol is due to the dissem-ination of the off-path values to individual nodes. To reduce this overhead, following the improvement proposed by Frikken et al. [89], we use a balanced commitment tree as an overlay on the physical aggregation tree. Due to space constraint, we do not discuss the details in this chapter. If a histogram with b buckets is considered, each off-path message is b times bigger than that in the Sum scheme, which makes the worst case node congestion in this protocol to be $O(b\Delta \log N)$.

5.4.3 Our Basic Protocol

We now describe our basic protocol to compute and verify an approximate Median. The basic protocol has two phases: Sampling phase, and histogram computation and verification phase. Below we discuss these phases in detail.

While collecting a sample of population values is highly energy-efficient compared to collecting all the values, we will later show that a sample can act as a good representative of the whole population. Also, we will show that only the sample size determines the performance of our algorithm, irrespective of the size of the population.

5.4.3.1 Sampling

In this phase, the BS collects a uniform sample of the sensed values from the network. To do so, the BS broadcasts the following message:

$$BS \rightarrow * : \langle SAMPLE, seed \rangle.$$

The sample request coming from the BS is broadcast in a hop-by-hop fashion and the nodes arrange themselves in a ring topology; nodes at the first hop from the BS belong to the first ring and so on. A node X considers the previous hop nodes as parents from which X has received the query message. Note that in the sampling phase, we do not use a tree topology, which is, however, used in the histogram computation and verification phase. We assume that there is a public hash function $F : \{ID, seed\} \rightarrow \{0, 1, \ldots, t - 1\}$, where ID represents the node id, $seed$ is the nonce broadcast during the query, and t is a positive integer which acts as a design parameter as discussed later. Each node, say X, hearing the query message applies the hash function $F(X, seed)$. If the resulting value is 0, then its sensed value, v_X, is considered to be one element in the sample. In that case, X computes $MAC(K_X, v_X)$ and sends the message $V_X = (X, v_X, MAC(K_X, v_X))$ to its parents. In addition to that, if X has child nodes, X also forwards the sample values and corresponding MACs received from the child nodes, say $V_{Z_1}, \ldots, V_{Z_c}$. The whole message from X looks as follows:

$$X \rightarrow Parents\,(X) : \langle V_X, V_{Z_1}, \ldots, V_{Z_c} \rangle.$$

When the BS receives all these messages, it verifies the corresponding MACs and outputs the list of values that are legal items of the sample. Note that the $seed$ is used in order to have different samples in different runs. Basically, the hash function is used to uniformly divide all of the nodes among t groups; the nodes belonging to the first group (i.e., output of the hash function is 0) are considered to constitute the sample. If the required sample size is S, one might set $t = N/S$. It is expected that this hash function uniformly maps N elements into t groups. To increase the chance that finally a sample of size no less than S will be collected, we could increase the number of groups from t to kt, and output the sample from more than k groups (e.g., $k + 1$ groups).

5.4.3.2 Histogram Computation and Verification

Once the BS obtains the sample, it sorts the items in ascending order. Then, the following steps are performed: (i) computing histogram boundaries, (ii) computing and verifying the buckets' count, and (iii) estimating the Median.

(i) **Computing Histogram Boundaries**.

We consider the number of buckets, b, as a parameter. In Sect. 5.5.2 we discuss how to choose this parameter. In this step, we equally divide the sample items into b buckets. We denote the buckets as $B_i = [q_i, q_{i+1}], 0 \leq i \leq b - 1$, where $q_0 = -\infty$, $q_i = E_{\lceil \frac{S}{b} \rceil i}$ and $q_b = +\infty$, as shown in Fig. 5.2. E_j represents the value of j-th item in the sample sorted according to the value, with j varying from 1 to S.

(ii) **Computing and verifying the buckets' counts**.

To compute the bucket counts, the BS and the sensor nodes run the histogram verification protocol described in Sect. 5.4.2. If there is no attack present in the network, at the end of this step the BS knows the number of nodes that belong to each bucket in the histogram. However, an attacker node can cause this verification to fail, and in that case, the protocol terminates returning a message, "attack detected". We discuss an attack-resilient solution in Sect. 5.6.

(iii) **Estimating the Median**.

Assuming that the verification in the previous step succeeds, we have the bucket counts $c_0, \ldots, c_{b-1}$ for the corresponding buckets. Our aim is now to find the bucket which contains the Median. In particular, we find j such that the following three constraints are satisfied:

$$e_l + c_0 + c_1 + \cdots + c_{j-1} < (N + 1)/2 \tag{5.2}$$
$$c_0 + c_1 + \cdots + c_j \geq (N + 1)/2 \tag{5.3}$$
$$c_j \leq 2\epsilon N \tag{5.4}$$

where e_l is equal to 0 in the first iteration and updated as follows in other cases. We first find j such that the first two in-equalities are satisfied. Then, we check if the above j also satisfies in-equality (5.4). Note that if in-equality (5.4) is satisfied, then

Fig. 5.2 Computing histogram boundaries: The histogram boundaries are computed using the sample collected in the previous phase

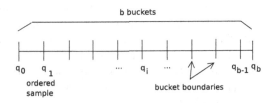

Fig. 5.3 Splitting the
bucket: If the bucket j,
which contains the Median
has more than $2\epsilon N$ elements,
the bucket is split in order to
meet ϵ approximation error
bound

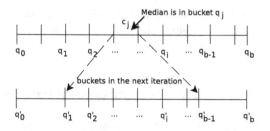

it is guaranteed that either q_j or q_{j+1} is ϵN away from the exact Median, which is
reported as our final estimate. If the in-equality (5.4) is not satisfied, we further split
j-th bucket equally into b sub-buckets. The new boundaries are updated as follows:
$q'_0 = q_0, q'_1 = q_j, \ldots, q'_{b-1} = q_{j+1}$, and $q'_b = q_b$. Bucket splitting is illustrated in
Fig. 5.3. The variable e_l is updated as $e_l = e_l + \sum_{i=0}^{j-1} c_i$. We iterate steps (ii) and (iii)
until the in-equality (5.4) is satisfied. During the above iteration, if we reach a point
where bucket j does not contain any sample items to split further, we stop returning
a message, "more sample items to be collected". We note that modifying the above
inequalities any other quantiles can be computed.

5.5 Security and Performance Analysis of Our Basic Protocol

5.5.1 Security Analysis

A node X which is selected in the sample sends an authentication code,
$MAC(K_X, v_X)$, to the BS so that the BS can authenticate the sensed value v_X, where
K_X is the pairwise key of X shared with the BS. An attacker node that is not legally
selected by the hash function cannot inject a false value in the sample without being
detected.

Moreover, because multipath routing scheme is used in the sampling phase, it is
highly likely that we will be able to collect a sample, even if a few compromised
nodes do not forward any messages. To establish the above observation, we consider
a simplistic scenario. Let us assume that there are w compromised nodes in total
and they are randomly distributed in the network. So, the probability of a randomly
selected node to be compromised is w/N, where N is the total number of nodes.
We also assume that each node has at least θ number of parents and the farthest
node is d hops away from the BS. We assume that unless all of the parents of a
node X are not compromised, X's message will reach the next hop—the probability
that this happens is $(1 - (w/N)^\theta)$. So, in the presence of the dropping attack by the
compromised nodes, the probability that a sample item finally reaches the BS is at
least $(1 - (w/N)^\theta)^d$. As an example, with $N = 1,000s, w = 50, \theta = 3$, and $d = 15$,
this probability is 0.998.

Like Chan et al's scheme, our histogram computation protocol is able to detect the *falsified sub-aggregate attack*, i.e., the attacker cannot modify the count of any bucket in the histogram without being detected. So, given that the verification succeeds, it is guaranteed that the final estimate is an ϵ-approximate Median.

5.5.2 Performance Analysis

In this section, we analyze the communication complexity of our basic protocol. In the first phase (i.e., during the sampling phase), the worst case node congestion occurs when a node (e.g., a node close to the BS) is required to forward all of the S samples coming from the network. So, the maximum node congestion in the sampling phase is $O(S)$. The cost of the second phase, which computes and verifies the histogram is $O(b\Delta logN)$, where b is the number of buckets, Δ is the degree of the aggregation tree, and N is number of nodes in the network. Note that our protocol iterates the second phase until the required approximation error bound is met. Our goal is to minimize the total cost of all iterations.

The second phase goes to the next iteration if the bucket b_j in which the Median lies contains more than $2\epsilon N$ population elements. We then further divide j-th bucket into b sub-buckets. We observe that further division is not possible if bucket j no longer contains a sample item, which is bound to happen within at most $log_b S$ iterations. If bucket j still contains more than $2\epsilon N$ population elements, we cannot do anything further but collect more sample items.

To make an estimate of the sample size, S, so that we do not need to perform an extra sampling phase in most of the cases, we present the following lemma.

Lemma 5.1 *The probability that more than pN population elements lie between two consecutive items of a sorted uniform sample of size S is $\phi(S, p) = (1 - p)^{S-1}$, where N is the population size.*

Proof Let A and B be two consecutive items in the sample after the sample items are sorted (as shown in Fig. 5.4). What we want to compute is the probability to have more than pN population elements between A and B. Once the sample item, A, is chosen, we have other $S - 1$ population elements remain to be chosen for the sample. To obtain the above probability, none of these $S - 1$ sample items should be chosen from the population interval which starts from A and is of length pN

Fig. 5.4 How far apart are two consecutive elements in the sample?

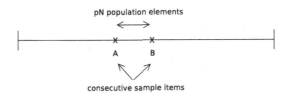

pN population elements

A   B

consecutive sample items

(i.e., the interval includes pN population elements). For each of these $S - 1$ sample items, the probability to be chosen not from that interval is $(1 - p)$. So, the probability that none of the $S - 1$ items will be there is $(1 - p)^{S-1}$. ◇

As an example, from Lemma 5.1, we see that $\phi(S, 2\epsilon) < 2.95 \times 10^{-5}$ for $S \geq 100$ and $\epsilon \geq 0.05$. This implies that if the user requires $\epsilon \geq 0.05$ and we use $b = 10$ buckets with $S = 100$, we require at most $log_b(S) = 2$ iterations to report the Median with probability $(1 - 2.95 \times 10^{-5}) \approx 1$. It is interesting to note that this result does not depend on the population size, N.

Now, to measure the trade-off between the number of buckets, b, and the number of iterations, which together determine the total cost of the algorithm, we present the following lemma.

Lemma 5.2 *The probability that more than $\gamma p N$ ($\gamma > 1, 0 < p < 1, \gamma p < 1$) population elements lie between the minimum and the maximum of pS consecutive sample items of a sorted sample of size S is*

$$\xi(S, p, \gamma) = \sum_{i=0}^{pS} \binom{S-1}{i} (\gamma p)^i (1 - \gamma p)^{S-1-i} \tag{5.5}$$

where N is the population size.

Proof Let A and B be the maximum and the minimum item among a subset of pS consecutive items in the sample while the sample items are sorted, as shown in Fig. 5.5. So, the expected number of population elements lying between A and B is pN. We would like to compute the probability to have more than $\gamma p N$ population elements lying between A and B, where $\gamma > 1$. Once the sample item, A is chosen, we have other $S - 1$ population elements remain to be chosen for the sample. To obtain the above probability, not more than pS items of these $S - 1$ sample items should be chosen from the population interval which starts from A and is of length $\gamma p N$ (i.e., the interval includes $\gamma p N$ population elements). For each of these $S - 1$ sample items, the probability to be chosen from that interval is γp. So, the probability that not more than pS items among the $S - 1$ items will be there is

$$\sum_{i=0}^{pS} \binom{S-1}{i} (\gamma p)^i (1 - \gamma p)^{S-1-i}.$$

◇

Fig. 5.5 What is the chance that $\gamma p N$ elements will fall within pS sample items, where $\gamma > 1$ and $0 < p < 1$?

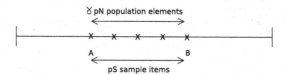

5.5.2.1 Number of Buckets Versus Number of Iterations

If we use $b = \frac{\gamma}{2\epsilon}$ buckets, which is of $O(\frac{1}{\epsilon})$, where γ is a constant greater than 1 and ϵ is the required error bound, then each bucket contains $\frac{2\epsilon}{\gamma} S$ sample items during the first iteration. So, the expected number of population elements in one bucket is $\frac{2\epsilon}{\gamma} N$. In Lemma 5.2, putting $p = \frac{2\epsilon}{\gamma}$, we can compute the probability that more than $\gamma \cdot \frac{2\epsilon}{\gamma} \cdot N = 2\epsilon N$ population elements fall in a bucket. As Expression (5.5) is a decreasing function of γ, by choosing the appropriate γ, we can make the above probability close to zero. As an example, for $\gamma = 2$, we observe that with sample size S such that $\epsilon S \geq 5$, (i.e., each bucket contains no less than 5 sample items in the first iteration) the above probability is less than 0.02 for all ϵ. That means, in this setting, our protocol ends in one iteration in 98 % cases. Finally, considering the cost of the histogram verification scheme, we see that the total cost of all iterations per node, when $b = O(\frac{1}{\epsilon})$, is $O(\frac{1}{\epsilon} \Delta \log N)$, where Δ is the degree of the aggregation tree.

On the other hand, if we use $b = O(1)$ buckets and equally divide the sample items in b buckets in each iteration, then, after $\log_b (\frac{\gamma}{2\epsilon})$ iterations, each bucket will contain no more than $\frac{2\epsilon}{\gamma} S$ sample items. So, as shown above, with the appropriate γ chosen, it is almost certain that our algorithm will end at this point. Thus, considering the cost to compute and verify the histogram in each iteration, the total cost of all iterations, when $b = O(1)$, is $O(\log_b \frac{1}{\epsilon} \cdot b \cdot \Delta \log N)$, where Δ is the degree of the aggregation tree.

5.6 Attack-Resilient Median Computation

Although our basic protocol, discussed in Sect. 5.4.3, detects *falsified sub-aggregate attack*, it fails to output an estimate of the Median in the presence of the attack. To address this problem, here we propose an extended approach so that we can compute an approximate Median even in the presence of a few compromised nodes.

We design the new approach based on the *divide and conquer* principle. We divide the network into several groups of nodes, which introduces resilience against the above attack. We run the verification algorithm individually for each group, which we call *intra-group verification*. Basically, we localize the attacker nodes to specific groups, i.e., we detect which groups are corrupted and which are not. Even if a few groups are corrupted, we still compute an estimate of the Median considering the valid groups. We do not assume that the groups have similar data distribution, which is the assumption exploited in other existing approaches such as SDAP [235] or RANBAR [29].

We may employ different grouping techniques based on node's geographic location or node IDs. We may also use grouping technique which is based on the nodes' positions on the aggregation tree. Once the group aggregate is computed, the group leader send it directly to the BS; to avoid having any node in the middle to drop group aggregates, we use a multipath routing mechanism. Due to space constraint, only geographical grouping technique is discussed here.

Also, we may exploit the robustness property of the Median computation to determine the maximum amount of error that can be injected by a given number of corrupted nodes, even if we do not perform the intra-group verification. In [195] we estimate this error while we leave it to the network user to fix the tradeoff between the error bound and the overhead due to intra-group verification.

5.6.1 Geographical Grouping

We assume that the BS has knowledge of the location of the nodes and each node knows its own location. The network is divided into several rectangular regions, where each region is identified by a pair of geographical points. The number of regions, g, and the location of the regions are selected considering a few factors. As one criterion, the regions might be chosen in such a way that an equal number of nodes belong to each group—if a region has lower node density, it is likely that it will be of larger geographical size. In addition, if the BS expects that a part of the network is more likely to be under attack, it may prefer to form smaller regions in that area to better localize the attacker. Finally, The g rectangular regions are specified by g pairs of diametrically opposite points, $(x1_i, y1_i)$, $(x2_i, y2_i)$, where $1 \leq i \leq g$. For each group i, BS also selects a node to be the group leader, GL_i. An example of this grouping is shown in Fig. 5.6.

Once the histogram boundaries are computed using the collected sample (as in our basic protocol), the BS initiates the histogram verification procedure. The BS sends a request to the corresponding group leaders with the necessary information to identify the regions. Receiving the request, a local aggregation tree is constructed which comprises of all of the nodes in the region with GL_i as the root. Then, the group histogram is computed locally and sent to the BS. If compromised nodes are

Fig. 5.6 Geographical grouping: In each region the group leader, GL_i, sends the region aggregate to the BS by multiple paths

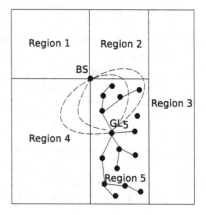

present in a few groups, the BS will be able to identify the corrupted groups. The BS accepts aggregates from only those regions, which passed the verification. The BS may further split the region which contains an attacker node and run the protocol again in the sub-regions. Eventually, this splitting can be iterated until the attacker node is identified or the percentage of verified values satisfies the BS (e.g., when the verified groups correspond to the 95 % of the nodes). Below we discuss the attack-resilient histogram computation and verification algorithm.

5.6.1.1 Algorithm Description

The nodes in each region locally perform the histogram computation and verification protocol described in Sect. 5.4.2 with the group leader acting as an agent of the BS in the corresponding group. To make the group leader GL_i an eligible agent of BS for group i, we need a few additional communication between GL_i and the BS. Below we focus on these additional messages skipping the detailed description of rest of the protocol, which can be found in Sect. 5.4.2. The messages exchanged between GL_i and the BS are authenticated using their pairwise key. To improve readability, we do not show these authentication fields in the messages below.

Query Dissemination

BS initiates the query by sending to each group leader GL_i via multiple paths the following message which contains the coordinates of the corresponding region:

$$BS \Longrightarrow GL_i : \langle (x1_i, y1_i), (x2_i, y2_i), GL_i \rangle.$$

In each region, the group leader, GL_i, broadcasts the received query message to its neighbour nodes, which again broadcast the same message, and so on. It is a scoped broadcast, i.e., if a node whose coordinate is outside of the corresponding region receives the message, it simply drops the message. During the query broadcast, a regional aggregation tree is formed with GL_i as the root, similarly as in the TAG [148] algorithm. The query message also contains required $\mu TESLA$ information (not shown above) so that each node in the region can authenticate the query.

After the query is disseminated, the nodes in each region locally perform the histogram computation and verification protocol described in Sect. 5.4.2.

Aggregation-Commit Phase

After the group leader GL_i receives the aggregated value from the nodes in group i, it forwards the following message to the BS:

$$GL_i \Longrightarrow BS : \langle GL_i, agg_i, commit_i \rangle,$$

where agg_i represents the computed histogram of group i, and $commit_i$ is the root of the *commitment tree* of region i.

Commitment-Dissemination Phase

The BS checks if the number of nodes in the computed histogram of the group is same as the total number of nodes in that group. If yes, it sends to GL_i the $\mu TESLA$ authentication information, $\mu T(commit_i)$. So, when GL_i broadcasts $commit_i$ in group i, each node can authenticate the message:

$$BS \implies GL_i : \langle GL_i, \mu T(commit_i)\rangle.$$

Result-Checking Phase

Each node checks if its value is incorporated in the computed histogram. If yes, node X sends a MAC over an "OK" message, $MAC(K_X, OK)$, which gets XOR-ed with other nodes' similar messages on their way to the group leader. Once GL_i receives the compressed OK message, say OK_i, from the nodes in its group, it forwards this message to the BS via multiple paths:

$$GL_i \implies BS : \langle GL_i, OK_i\rangle.$$

As the BS knows which nodes belong to which group, it can verify OK_i messages and hence can identify valid group aggregates.

5.6.1.2 Security Analysis

We recall from section that the histogram computation and verification protocol, when executed on the whole network, can detect if there is any falsified sub-aggregate attack. That means, if a malicious node X fabricates the histogram of its sub-tree or if X simply does not participate in the protocol, the BS can detect the attack and flags that the computed histogram is corrupted. Our intra-group verification protocol is different from the basic one only in the following aspects: (i) the histogram of the whole network is considered as the aggregate of the group histograms and each group histogram is computed and verified individually, (ii) the group leader, GL_i exchanges a few messages with the BS, discussed in Sect. 5.6.1.1, which enable GL_i to play the role of BS in group i.

The messages exchanged between GL_i and the BS are routed via multi-paths so that they reach the destination even if an attacker node in the middle drops these messages. The communication between GL_i and the BS is also authenticated with their pairwise key. Moreover, GL_i receives from the BS the $\mu Tesla$ authentication information for the messages which are to be broadcast in the group, e.g., the query message and the $commit_i$ message. So, assuming a node X knows its location, X can securely determine to which group it belongs and the ID of the group leader, and X can also authenticate the query and the $commit_i$ message endorsed by the BS.

After the BS receives the group histogram from group i, (i.e., the agg_i message) the BS verifies if the number of nodes reflected in the group histogram is same as the number of nodes in the group. Also, after receiving the OK_i message from group

i, the BS verifies if this message correctly represents, in compressed form, the OK message of all the nodes in group i. The above two checks enable the BS to correctly identify the corrupted groups, if any.

5.6.1.3 Performance Analysis

On average, the number of nodes in one group is $N' = \frac{N}{g}$, where the network is divided into g groups. So, the worst case node congestion inside one group for running the histogram verification algorithm is $O(b \cdot \Delta \cdot \log N')$, where b is the number of buckets in the histogram and Δ is the number of neighbours of a node on the aggregation tree. Considering the analysis given in Sect. 5.5.2.1, with $b = O(\frac{1}{\epsilon})$, the worst case communication overhead per node is $O(\frac{1}{\epsilon} \cdot \Delta \cdot \log N')$. In addition, a node needs to forward the messages exchanged between the group leaders and the BS, which is of $O(g)$ communication overhead in the worst case.

5.7 Simulation Results

In this section, we report on a simulation study that examined the performance of our basic protocol discussed in Sect. 5.4. Recall that, in the first phase, we collect a sample of sensed values from the network, and the performance of the rest of the protocol depends on the quality of this sample. The goal of the simulation experiments reported below is to study the impact of the sample on the overall performance of the Median computation protocol. In particular, we verify the results we obtained via analysis, in Sect. 5.5.2, about the inter-relationship among parameters, such as error bound ϵ, sample size S, and the number of buckets b in the histogram.

Through simulation we do not evaluate the overhead of in-network communications in our protocol. The analytical results on the communication overhead of the sampling phase and the histogram computation and verification phase are discussed in Sect. 5.5.2.

5.7.1 Simulation Environment

In our basic setup, the network size is 1,000 nodes. We also vary the network size to show that it does not have a significant impact on our sampling-based approach. In our simulation, the typical value we take for the ϵ error bound varies from 5 to 15%. Each node has one sensed value, while our goal is to compute an approximate Median. We use the method of independent replications as our simulation methodology. Each simulation experiment was repeated no less than 1,000 times with different seeds.

5.7.2 Results and Discussion

Here, we discuss the results obtained in our simulations. We observe that 95 % confidence interval of all the quantities on the following plots are within 5 % of the reported value.

What is the chance that one sampling phase is not enough? In Lemma 5.1, we analytically computes this probability which we evaluate via simulation here. For each pair (S, ϵ), we collect a sample of size S and we compute the number of time, τ there are more than $2\epsilon N$ elements between the two consecutive sample items containing the Median. The total number of runs performed is 1,000,000. The resulting $\phi'(S, 2\epsilon)$, which is the observed approximation of $\phi(S, 2\epsilon)$, is plotted in Fig. 5.7. It is worth noticing that the value of $\phi'(S, 2\epsilon)$ is less than 4×10^{-5} for $\epsilon > 0.05$ when the sample size S is more than 95. In fact, as expected, for a given ϵ, an increase of the value of S decreases $\phi'(S, 2\epsilon)$. Finally, we verify that $\phi'(S, 2\epsilon)$ does not change significantly (not shown in the figure) even if the population size, N, is bigger.

Number of buckets versus Number of iterations. In Sect. 5.5.2, we analyzed the dependence of the number of iterations of our algorithm on the number of buckets chosen, which we validate here via simulations. First, we estimate the number of buckets required to end our protocol in one iteration in most cases. Figure 5.8a illustrates the % of cases our protocol ends in the first iteration. The figure confirms our analysis that, for considering $\gamma = 2$, if we use more than $\frac{1}{\epsilon}$ buckets (i.e., 20, 10, 7 buckets for $\epsilon = 0.05, 0.10, 0.15$, respectively), it is highly likely that we need just one iteration. Finally, Fig. 5.8b shows the average number of iterations required

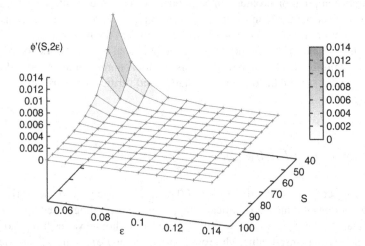

Fig. 5.7 Computing the chance that we need to collect more sample items: Given an ϵ, we choose a sample size so that the probability that we need to redo the sampling is close to zero

Fig. 5.8 The number of iterations versus the number of buckets: If the number of buckets is $O(\frac{1}{\epsilon})$, it is highly likely that our algorithm ends in one iteration. **a** % times ending in one iteration. **b** Average number of iterations

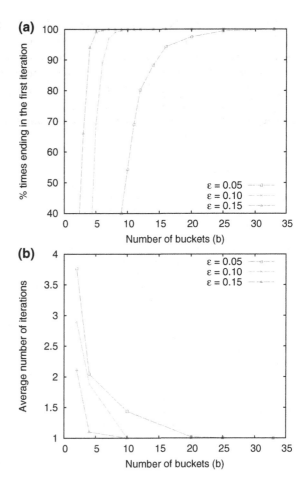

using different number of buckets, where $\epsilon = 0.05$ and $S = 100$. This validates our analysis that the average number of iterations is $O(log_b(\frac{1}{\epsilon}))$ when b buckets are used.

5.8 Concluding Remarks

While researchers already addressed the problem of securely computing aggregates such as Sum, Count, and Average, to the best of our knowledge, there is no prior work on secure computation of the Median. However, it is widely considered that the Median is an important aggregate. In this chapter, we proposed a protocol to compute an approximate Median and verify if it is falsified by an attack. Once the protocol is executed, the base station either possesses a valid approximate Median or it has detected an attack. Further, we proposed an attack-resilient algorithm to compute the Median even in the presence of a few compromised nodes. The evaluation via both analysis and simulation shows that our approach is efficient and secure.

This chapter dealt with an attack on the data aggregation protocol: We considered an attacker that aims to exploit the aggregation protocol to let the collecting node accept a false aggregate. The next chapter deals with another security aspect related to data aggregation: The privacy of the nodes participating in the aggregation procedure.

Chapter 6
Privacy in Data Aggregation

We started addressing the security of in-network data aggregation in the previous chapter, where we addressed the problem of the Median computation where attacker nodes can be present. In this chapter we address another security problem related to the data aggregation: The node's privacy. In fact, in the data aggregation technique, some sensor nodes need to send their individual sensed values to an aggregator node, empowered with the capability to decrypt the received data to perform a partial aggregation. This scenario raises privacy concerns in applications like personal health care and the military surveillance.

The contributions of this chapter are two-fold: first, we design a private data aggregation protocol that does not leak individual sensed values during the data aggregation process. In particular, neither the base station nor the other nodes are able to compromise the privacy of an individual node's sensed value. Second, the proposed protocol is robust to data-loss; if there is a node-failure or communication failure, the protocol is still able to compute the aggregate and to report to the base station the number of nodes that participated in the aggregation. To the best of our knowledge, our scheme is the first one that efficiently addresses the above issues all at once.

6.1 Introduction

Inspired by the low-cost, flexibility and ubiquitousness of this technology, researchers [111] are envisioning sophisticated applications of WSNs including sensors being installed in personal environment, such as houses, and human body. However, one issue raised by this range of applications is the privacy of the data being collected. In [111], a future application is cited, which involves sensing power or water usage of private households to compute the average trend of a region. People might not

© Springer Science+Business Media New York 2016
M. Conti, *Secure Wireless Sensor Networks*, Advances in Information Security 65,
DOI 10.1007/978-1-4939-3460-7_6

agree to allow these applications to intrude their personal world if the privacy of the collected information is not protected. The goal of this chapter is to design a scalable, efficient, data-loss resilient, privacy-preserving data aggregation algorithm for WSNs.

To achieve this goal, one might suggest to adapt the existing privacy-preserving algorithms designed for data mining applications [2,87], but, unfortunately, these algorithms are too computational expensive to meet the severe resource constraint of sensor nodes. Exploring alternative paths, researchers [32, 96, 111, 134, 154, 165, 239, 242] have presented some proposals to solve the privacy problem in data aggregation. The goal of this chapter is to consider the following three requirements at the same time:

1. To prevent the sensed value of an individual node from being disclosed to other nodes during the aggregation process.
2. To prevent the sensed value of an individual node from being disclosed to the BS, i.e., the BS will have access only to the data aggregate.
3. The possibility of a node becoming off-line during the aggregation process, or a message being lost before reaching the BS, should not affect the correctness of the aggregate computed based on the nodes that participated in the aggregation.

The main idea behind the design of our algorithm is as follows. The nodes in the network divide themselves into clusters. The aggregate of the nodes within a cluster is computed in such a way that no individual sensor reading is leaked during this process. To *obfuscate* the individual sensor readings, we make use of *twin-keys* shared by node pairs within a cluster, which are established in an onetime set-up phase. After the cluster aggregates are computed, they are sent in clear text to be further aggregated to compute the final aggregate—usually via a tree-based aggregation algorithm.

The work presented in this chapter is the first one, to the best of our knowledge, that achieves both the following properties: First, it provides a mechanism that preserves the privacy of the data contributed by a sensor to the aggregate value. That is, the individual values as well as the identity of the contributing nodes cannot be derived by any node in the network, as well as by the BS. Second, the protocol is robust against communication and node failures. In Table 6.2 we summarize the features of our proposed protocol compared with other protocols in literature. Moreover, our recent work on external and internal threats is presented in [240], and further investigation on privacy-preserving data-collection for smart metering can be found in [7].

Organization

The rest of the chapter is organized as follows. We present the related work in Sect. 6.2. Section 6.3 presents the assumptions and the threat model considered in this chapter. In Sect. 6.4 we give the overview of our proposed solution, while a detailed presentation is discussed in Sects. 6.5 and 6.6. In Sect. 6.7, we present the security and performance analysis of our protocol. We finally conclude in Sect. 6.8.

6.2 Related Work

Researchers [91, 148] proposed in-network aggregation algorithms which fuse the sensed information en route to the BS to reduce the communication overhead. Several algorithms are designed to compute aggregates such as Count, Sum, and Average.

The research community also examined a few security issues related to aggregation algorithms. Wagner [220] addressed the problem of resilient data aggregation in the presence of false data injection attack by a few malicious nodes, and provided guidelines for selecting appropriate aggregation functions in a sensor network. Yang et al. [235] proposed SDAP, a secure hop-by-hop data aggregation protocol using a tree-based topology to compute the correct Average in the presence of a few compromised nodes. Chan et al. [38] designed a novel verification algorithm by which the BS could detect if the computed aggregate was falsified. Another approach for computing Count and Sum, even if a few compromised nodes inject false values, was proposed by Roy et al. [195]. However, none of the above algorithms address the privacy issues of data aggregation.

There exists a body of work that addresses privacy issues in data mining applications. In [2, 87], the authors proposed data perturbation techniques to protect the private values, whereas a few secure multi-party computation schemes were designed in [63, 108, 236]. However, these privacy-preserving algorithms are too much computation expensive to be applicable for the low-end nodes in a sensor network.

Privacy Homomorphism (PH) proposed by Rivest et al [187] allows to aggregate encrypted data. PH is an encryption transformation that enjoys the following property, related to an operation "∘". Given an encryption function, $E : S \times R_1 \rightarrow R_2$, and a decryption function, $D : S \times R_2 \rightarrow R_1$, where R_1, R_2 are rings and S is the key-space, for $a, b \in R_1$ and $s \in S$, the following equation holds: $a \circ b = D_s(E_s(a) \circ E_s(b))$. Domingo-Ferrer [78] proposed a PH that preserves both addition and multiplication ("∘" is "+" and "·", respectively). The proposal has been proven to be secure against known-cleartext attack (as long as the ciphertext space is much larger than the cleartext space).

In Girao et al.'s work [96], the PH is used to allow the aggregator node to compute the correct encrypted aggregate from the encrypted values coming from sensor nodes. This allows the protocol to guarantee the privacy of the sensor nodes against a passive eavesdropper. However, as all of the nodes share the same encryption key with the base station, the protocol does not guarantee the privacy of individual sensed data against other nodes or the BS.

In [32], the authors propose a solution for data aggregation that protects the privacy against other nodes. The authors assume that each node n_i shares a key k_i with the BS. Basically, each node n_i adds a random number to its sensed value where the random number is determined by k_i. After receiving the encrypted aggregate, the BS filters out the correct aggregate by subtracting all the random numbers added by the nodes. We observe that this scheme does not protect privacy of individual sensed values if the BS eavesdrops over the network. Moreover, this scheme is critically vulnerable to message loss, which is very common in a WSN. If just one message is lost, the

BS obtains a bogus aggregate, without suspecting any problem. The author propose how to cope with this last issue, adding the list of contributing nodes or the list of nodes that did not contribute (whichever is the shorter one). However, note that this solution does not prevent this list from being $O(N)$ in length, where N is the number of sensors in the network.

In [111], the authors propose two different solutions: CPDA and SMART. The former gives a solution for data aggregation preserving node-privacy against other nodes. We observe that CPDA can be extended to provide privacy against BS and furnish a solution against data-loss as well. However, as stated by the authors, the overhead of this protocol is high. Indeed, they use the anonymization sets, where each node out of the C nodes within a cluster has to send (and receive) $C - 1$ messages, resulting in $O(C^2)$ sent (and received) messages within each cluster, for each aggregation phase. Furthermore, each single node has to encrypt and decrypt $O(C)$ messages, and the cluster head has to compute the inverse of a $C \times C$ matrix, for each aggregation phase. The latter proposal, SMART, is more efficient than the first proposal. However, it does not protect privacy against the BS, and suffers from the same problem of message-loss as in [32].

Researchers also looked into source privacy issues in sensor network. In [234], Yang et al. prevent a global adversary from identifying a node as the event source.

6.3 Network Assumptions and Threat Model

In this section, we describe the network assumptions and the threat model considered in the rest of this chapter. We consider a static multihop WSN of N sensor nodes and a single base station (BS). We consider sensor nodes similar to the current generation of sensors (e.g., Berkeley MICA2 Motes [119]) in their computational and communication capabilities and power resources, while the BS is a laptop-class device supplied with long-lasting power.

A pair-wise key mechanism, like the ones in [47, 75, 85, 144], is used to enable secure communications among the network nodes.Our protocol is independent from the particular mechanism used. We further assume that a set of K keys (key-ring), is pre-loaded in each node, using the set-up procedure of Eschenauer and Gligor's protocol [85]: the K keys are randomly chosen from a larger key-pool of size P. BS does not know any information about this pool. Note that this set of keys is not related to the pair-wise key establishment while it is exclusively used in our "twin-key" establishment protocol.

In our proposal, we use a clustering mechanism to group the nodes in several clusters. It is out of the scope of this chapter to give a detailed description of the cluster formation algorithm. Our protocol makes use of a cluster formation algorithm such as the one in [41]. The basic building block for cluster formation in [41] is as follows: each node applies an hash function, $H : (seed|x) \rightarrow y \in [1..deg]$, where deg is the average degree of a node and x is a node ID. The node having the largest result among neighbors (y_{max}) becomes the *leader* and announces its status.

Table 6.1 Privacy in data aggregation: Notations

Symbol	Meaning
P	Key Pool;
K	Size of the node key-ring;
$\mathcal{K}_{ID}$	Key-ring set of the node ID;
e	Indicates the id of the executing node in the procedures;
k_i	ith key in the key pool for twin-key;
C	Number of nodes in each cluster;
A	The number of twin-keys each node needs to establish;
V	The number of alive twin-keys required for active participation;
R	The size of message in terms of declared keys;
r	The number of twin-keys each node can initiate in each round;
H	A hash function;
d_{ID}	Sensed (private) value of the node ID;

Nodes with value smaller than y_{max} wait to hear from a leader to set that node as the cluster head. Special cases, conflict resolution and the security of this protocol are discussed in [41].

In this chapter, we focus on in-network computation of the Sum aggregate. Note that it is possible to extend the Sum aggregate to implement other aggregation function as well, such as Count and Average.

We assume that the aim of the adversary is to compromise the privacy of a node. In particular, it will not drop or modify messages if this does not help him to violate the privacy of a node. We consider the adversary to be able to:

1. Eavesdrop all of the network communications;
2. Control a fraction of nodes;
3. Obtain any information from the BS.

Table 6.1 summarizes the notation used in this chapter.

6.4 Protocol Overview

The key elements of our protocol are the following: first, we establish *twin-keys* for different pairs of nodes in the network. We require the twin-key establishment to be anonymous—each node in a pair cannot derive the identity of the other node (*twin-node*) it is sharing a twin-key with. Second, for each aggregation phase, we use an anonymous liveness announcement protocol to declare the liveness of each twin-key—each node becomes aware of whether a twin-key it possesses will be used by the anonymous twin-node. Finally, during the aggregation phase, each node encrypts its own value by adding *shadow values* computed from the alive twin-keys it holds.

As a result, the contribution of the shadow values for each twin-key will cancel out each other.

Our protocol consists of three major steps as follows:

1. Local cluster formation: In this step, we require the network nodes to group themselves into clusters. In literature, there are different solutions for the cluster formation. The description of a detailed cluster formation algorithm is out of the scope of this chapter. In the following, we assume that a cluster algorithm such as the one proposed in [41] is used. For ease of exposition, from now on we also assume that each cluster has a fixed number of nodes, C. In the following steps of the protocol, we further require each cluster to form different logical Hamiltonian circuits. Note that an extension of the cluster formation in [41] can be used for the Hamiltonian circuit formation. As an example, each node, say n_i, can start setting a new circuit by randomly choosing one cluster node, say n_j, as its right neighbor in a circuit. Node n_i communicates its choice to node n_j. The selected node, n_j, will choose its own right neighbor within the nodes in the same cluster, that are not yet selected. Eventually, the last node will select node n_i as its right neighbor, to complete the Hamiltonian circuit. We require for each pair of nodes that are neighbors in the circuit to share a pair-wise key.

2. Twin-key establishment: This step is performed independently within each local cluster. We recall that we assume that each node contains a pre-deployed key-ring of K symmetric keys, randomly chosen from a larger common key-pool of size P. In this step, each node n_i anonymously checks which ones of its K keys are also shared with other nodes in the same cluster. In particular, a node is required to have at least A out if its K keys shared within its local cluster. This step will be further discussed in Sect. 6.5.

3. Data aggregation: This is the actual aggregation step of our protocol. Note that, other than this step, all the previous steps are performed only once during the set-up phase. The data aggregation step can be further divided into two main parts:

    3.1. First, each cluster computes the aggregated value of its nodes, together with a twin-key liveness announcement procedure. During this phase an aggregate is routed twice along the Hamiltonian circuit. Each node adds to the aggregate its own sensed value. At the same time, for each alive twin-key it adds (or removes, in accordance with the liveness announcement) a corresponding shadow value. As a result, the cluster head obtains the correct aggregate for the cluster.

        The liveness announcement guarantees that any shadow value, computed from a twin-key, that is added in the aggregation by one node, will be removed by another node that shares the same twin-key.

    3.2. At the end of step 3.1, there will be several nodes in the network that acted as cluster heads. These nodes own the corresponding cluster aggregates. Now, we want to further aggregate all of these values and to pass the final aggregate to the BS. In this step, we use a tree-aggregation hierarchical structure, commonly discussed in literature. In particular, we assume to use

Fig. 6.1 Aggregating the
individual cluster aggregates:
A tree aggregation protocol
is used to compute the final
aggregate

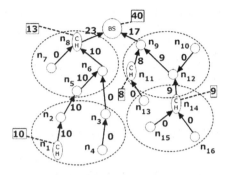

the TAG algorithm [148] with the following modification. The cluster head
nodes will contribute to the aggregate with the cluster aggregate computed
in step 3.1. All of the other nodes do not contribute to the aggregate while
they forward the aggregate computed from the received sub-aggregates. As
a result, the BS will receive the sum of the values owned by all of the cluster
heads.

Figure 6.1 illustrates this procedure with an example. The cluster heads n_1,
n_8, n_{11} and n_{14} possess the corresponding cluster aggregates 10, 13, 8 and
9, respectively. All of the other nodes contribute to the final aggregate with
value 0. Finally, the BS will obtain the aggregate result 40.

We discuss the step 3.1 in Sect. 6.6, while we refer to [148] for the TAG tree-
aggregation of the step 3.2.

6.5 Twin-Key Agreement

In this section we describe the set-up phase. The aim of this step is, for each node,
to establish a number of twin-keys with other nodes. In particular, we say that node
n_i established a twin-key with another node (twin-node) in the cluster if n_i is aware
of the fact that there is a node in the cluster sharing a key with it. Note that n_i does
not know the identity of its twin-node. The requirements of a twin-key establishment
are:

- The twin-keys are only known to the owners. They cannot be eavesdropped by
  other nodes or by the outside attackers.
- The twin-nodes (nodes that agree on a twin-key) cannot determine each other's
  identity, i.e., the twin key is established anonymously.

Furthermore, to improve on the level of anonymity, we require each node to
establish at least a given number, A, of twin-keys.

The twin-key agreement is a relay-based protocol. The twin-keys are initiated by
nodes, passed through the circuit of the local cluster, and accepted by other nodes.

Our protocol assumes that a key-ring of symmetric keys are pre-deployed in each node using the same approach as in Eschenauer et al.'s scheme [85]. For each sensor node, the manufacturer stores in the node's memory K keys randomly selected from a pool of $P \gg K$ symmetric keys. As a result, two nodes in the same cluster will share a given key with a probability depending on K and P as studied in [76].

6.5.1 Twin-Key Agreement: Protocol Description

At the beginning of the twin-key agreement protocol, each node runs PREPARE procedure (Algorithm 9). Variable a is used to keep track of the number of twin-keys that the node still needs to establish. *List* is a list of $< seed, key >$ pairs used to keep track of the twin-keys waiting to be agreed by other nodes, where each key is identified by a seed which is a random number. *TKList* is a list of already established twin-keys. $\mathcal{K}'$ is the list of keys not yet used for twin-key agreement. Finally, *Valid* indicates whether the executing node will participate in the subsequent aggregation procedure.

---

1 **begin**
2     $a = A$ //Number of Twin-key needs to be agreed ;
3     $List = empty$ //Seed-Key pairs sent out ;
4     $TKList = empty$ //Agreed Twin-keys ;
5     $\mathcal{K}' = \mathcal{K}_e$ //Twin-keys can be initiated ;
6     $Valid = true$ //Will participate in aggregation ;
7 **end**

Algorithm 9: PREPARE

PREPARE procedure is executed by each single node. After this procedure is carried out, each cluster head executes procedure INITIATEAGREEMENT (Algorithm 10). First, the CH creates a message M with R empty seed-key pairs (line 2), where R is a design parameter. The message format is: $\langle S, \langle s_1, h_1 \rangle, \ldots, \langle s_R, h_R \rangle \rangle$, where S is the total number of twin-keys to be established, and $\langle s_i, h_i \rangle$, $1 \leq i \leq R$, denote the twin-keys declared in the message waiting to be agreed on. The cluster head initializes the message by setting $S = A \cdot C$. This is used to meet the requirement that each of the C nodes in the cluster shares at least A twin-keys.

After generating M, CH randomly selects r positions out of the R ones in M (line 3) and randomly selects and remove r keys from $\mathcal{K}'$ (line 4). Then, for each selected key, it generates the pair $< s_i, H(k_i) >$ (line 5), where $H(k_i)$ is the hash of k_i and s_i is a random number associated with k_i. These pairs are copied in the r selected positions of M. Note that the r positions are randomly selected (line 3) in order to prevent the attacker to associate a node identity with a given position in M. Finally, CH sends M to one of its neighbor. This implicitly determines the direction of the message in the circuit.

```
1  begin
2  |   M ← ⟨C × A, ⟨0, _, ⟩₁, ..., ⟨0, _⟩_R⟩ ;
3  |   randomly select i₁, ..., i_r from {1, 2, ... R} ;
4  |   randomly select and remove k₁, ..., k_r from 𝒦' ;
5  |   randomly select key seeds (random number) s₁, ..., s_r ;
6  |   List = List ∪ {⟨s₁, k₁⟩, ..., ⟨s_r, k_r⟩} ;
7  |   replace ⟨0, _⟩_{i₁}, ..., ⟨0, _⟩_{i_r} with ⟨s₁, H(k₁)⟩_{i₁}, ..., ⟨s_r, H(k_r)⟩_{i_r} in M ;
8  |   send M to the next node //encrypted with pair-wise key ;
9  end
```

Algorithm 10: INITIATEAGREEMENT

Each node that receives the twin-key agreement message executes procedure RECEIVEMESSAGE(M) (Algorithm 11). This procedure consists of the following main steps performed by each node n_i:

(i) n_i checks the newly agreed keys which it had proposed before (lines from 3 to 16). That is, for each key declared in the previous round (line 4), n_i checks whether the key has been accepted. In particular, if the declaration of the key is still in the message M (line 6) it means that no other node agreed for that key. Otherwise, the key will be considered shared with someone (line 9). If the executing node has not yet established enough twin-keys (i.e. $a < A$), the newly agreed keys will be counted in the node's number of agreed keys, a, and the cluster's number of agreed keys, S. We recall that the declaration of a key is a pair $< s_i, H(k_i) >$, where s_i is a random number used to keep track of the particular instance of the key held by the node that declare that key. This is used in order to avoid confusion if the same key is proposed by more than two nodes in the cluster. As a result, if a node n_i declares key k_1 in M and k_1 is removed by n_j, a third node n_z re-declaring the same key, k_1, will use a different seed. If the message goes back to n_i, it can indeed understand that its key has been accepted and that the one declared in the message is just another instance of k_1 declared by some other node.

(ii) n_i checks for twin-keys proposed by other nodes that it can agree on (line from 17 to 29). For all keys' hashes in M (line 17), it checks if it has the corresponding key (line 18). If n_i never agreed on that key (line 19) it agrees on this key (line 20 and 21). Also, the corresponding counter is updated, if necessary (lines from 22 to 25). In any condition, when the key is agreed, the corresponding declaration is removed from the message M.

(iii) n_i proposes new keys to be agreed on (lines from 30 to 42). If the number of twin-keys established by n_i are not yet enough (line 30), it checks if it still has some keys to propose for agreement (line 31). If this condition is not satisfied, then the node will not participate in the aggregation phase (lines 32), because its privacy is not protected enough. In this case, the node also updates the variable S, in order to allow the CH to end the protocol without its own participation (line 33). If the node has other keys it did not try to share yet (i.e. condition in line 31 does not hold) it selects available keys from $\mathcal{K}'$ and declares these keys in the message M. The number of keys that can be declared in M will be bounded by $t = min\{t, r, |\mathcal{K}'|\}$—$t$ is the number of free positions in M (line 35), and r is the maximum number of keys declared in each round.

(iv) n_i sends the message or concludes the protocol (lines from 43 to 51). In particular, the only node that can end the protocol is the CH. The cluster head can terminate the protocol only if each node in the cluster either agreed on A keys (lines 11–12 and 23–24) or refused to participate in the aggregation (line 33). In all of the other cases, that is if $S \neq 0$ or the executing node is not the CH, it just sends the message to its neighbor.

An example of twin-keys agreement is shown in Fig. 6.2. In this example, the CH (node n_1) initiates the protocol by sending message 1, where it proposes two keys (k_1, k_3) for the agreement. When the announcement of k_1, that is $< s_1, H(k_1) >$, reaches n_4, it agrees on this key and removes the announcement from the message. Eventually, when n_1 receives message 6, it knows that k_1 has been agreed with someone.

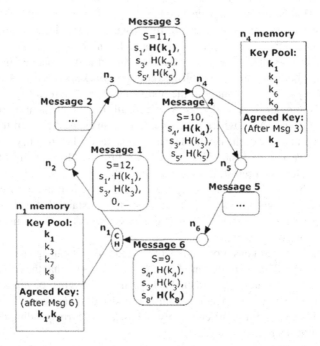

Fig. 6.2 Twin-key agreement example

```
 1 begin
 2 │   ⟨S, ⟨s₁, h₁⟩₁, . . . , ⟨s_R, h_R⟩_R⟩ ← M ;
 3 │   if !EMPTY(List) then
 4 │   │   for all ⟨s′, key′⟩ ∈ List do
 5 │   │   │   remove ⟨s′, key′⟩ from List ;
 6 │   │   │   if ∃sᵢ = s′ then
 7 │   │   │   │   replace ⟨sᵢ, hᵢ⟩ᵢ with ⟨0, _⟩ᵢ in M ;
 8 │   │   │   else
 9 │   │   │   │   TKList = TKList ∪ {key′} ;
10 │   │   │   │   if a > 0 then
11 │   │   │   │   │   a = a − 1 ;
12 │   │   │   │   │   replace S with S − 1 in M ;
13 │   │   │   │   end
14 │   │   │   end
15 │   │   end
16 │   end
17 │   for all i, i = 1 . . . R, sᵢ ≠ 0 do
18 │   │   if ∃key′ ∈ 𝒦_e, H(key′) = hᵢ then
19 │   │   │   if key′ ∈ 𝒦′ then
20 │   │   │   │   TKList = TKList ∪ {key′} ;
21 │   │   │   │   remove key′ from 𝒦′ ;
22 │   │   │   │   if a > 0 then
23 │   │   │   │   │   a = a − 1 ;
24 │   │   │   │   │   replace S with S − 1 in M ;
25 │   │   │   │   end
26 │   │   │   │   replace ⟨sᵢ, hᵢ⟩ᵢ with ⟨0, _⟩ᵢ in M ;
27 │   │   │   end
28 │   │   end
29 │   end
30 │   if a > 0 then
31 │   │   if 𝒦′ = φ then
32 │   │   │   Valid = false ;
33 │   │   │   replace S with S − a in M ;
34 │   │   else
35 │   │   │   t = number of sᵢ = 0 in M ;
36 │   │   │   t = min{t, r, |𝒦′|}; randomly select and remove i₁, . . . , iₜ from {1, 2, . . . R} ;
37 │   │   │   randomly remove k₁, . . . , kₜ from 𝒦′ ;
38 │   │   │   randomly select key seeds (random number) s₁, . . . , sₜ ;
39 │   │   │   List = List ∪ {⟨s₁, k₁⟩, . . . , ⟨sₜ, kₜ⟩} ;
40 │   │   │   replace ⟨0, _⟩_{i₁}, . . . , ⟨0, _⟩_{iₜ} with ⟨s₁, H(k₁)⟩, . . . , ⟨sₜ, H(kₜ)⟩_{iₜ} in M ;
41 │   │   end
42 │   end
43 │   if executing is CH then
44 │   │   if S = 0 then
45 │   │   │   broadcast twin-key agreement over ;
46 │   │   else
47 │   │   │   send M to the next node //encrypted with pair-wise key ;
48 │   │   end
49 │   else
50 │   │   send M to the next node; //encrypted with pair-wise key ;
51 │   end
52 end
```

Algorithm 11: RECEIVEMESSAGE(M)

6.6 Data Aggregation

In this section we explain the data aggregation phase. In particular, for ease of exposition, we describe it in two consecutive steps:

 (i) Twin-keys liveness announcement;
(ii) Data aggregation with shadow values.

Despite we present them separately, these two procedures can run together as discussed at the end of this section.

In the liveness announcement procedure, all of the nodes anonymously declare the liveness of the twin-key they posses. Each node will check whether the number of currently alive twin-keys is enough to protect the privacy of the sensed value. With $V \leq A$ we indicate the number of twin-keys that a node requires to be used during the aggregation in order to reach a satisfactory level of privacy. Assume that at least V of the node's twin-key are announced alive. Then, in the aggregation phase a node will add to the aggregate value computed so far its private value and the sum of the shadow values computed based on its alive twin-keys. However, if less than V twin-keys are announced as alive by the corresponding twin nodes, the node will add to the aggregate only the shadow values without its own private value.

In Sect. 6.6.3, we show how to combine the twin-key liveness announcement and the aggregation together.

6.6.1 Twin-Key Liveness Announcement: Protocol Description

In the twin-key liveness announcement, each node first executes the procedure ANNOUNCEPREPARE (Algorithm 12). $ATKList^+$ and $ATKList^-$ are used to record the alive twin-keys. Twin-keys in $ATKList^+$ will be used to compute positive shadow values and twin-keys in $ATKList^-$ will be used to compute negative shadow values. The data-type of $ATKList^+$ and $ATKList^-$ are *Set* and *Bag* respectively: A key appears at most once in $ATKList^+$ but could appears more than once in $ATKList^-$. Finally, *Avalid* is used to record whether the executing node will participate in the data aggregation.

```
1 begin
2 │   ATKList⁺ = φ ;
3 │   ATKList⁻ = φ ;
4 │   TempATKList⁺ = φ ;
5 │   Avalid = true ;
6 end
```

Algorithm 12: ANNOUNCEPREPARE

After the ANNOUNCEPREPARE procedure has been executed by all of the nodes, CH initiates the twin-key liveness announcement, i.e. executes the procedure CH_ANNOUNCE (Algorithm 13). It initializes the number of participating node, $T = 0$ (line 2). Then, it will consider A of its twin-keys (line 3). For each key, k_i, it will generate a random seed, s_i (line 4), and it will write all the pairs $< s_i, H(k_i) >$ in the message it sends to its neighbor (line 7).

```
1 begin
2 │   T = 0 ;
3 │   Select A twin-keys from TKList: k₁, ..., kₐ ;
4 │   Select A random numbers: s₁, ..., sₐ ;
5 │   TempATKList = {(s₁, k₁), ..., (sₐ, kₐ)} ;
6 │   List = {(s₁, H(k₁)), ..., (sₐ, H(kₐ))} ;
7 │   send (T, List) to the next node //encrypted with pair-wise key ;
8 end
```

Algorithm 13: CH_ANNOUNCE

Each node, except the CH, receiving the liveness announcement message for the first time executes the procedure FIRSTROUND (Algorithm 14). The CH never executes this procedure. In particular, for each h_i in the *List* within the message M (line 4) it checks if it has a twin-key, k, such that $H(k) = h_i$ (line 5). If this is the case it adds the key, k, in its list *ATKList⁻* (line 6) and removes the corresponding key announcement from the message (line 7). Furthermore, the node declares the liveness of all its A' other keys not yet known as alive, where $A' = A - |ATKList⁻|$ (line 10). To do so, as for the Procedure CH_ANNOUNCE, it selects a random seeds s_i for each key k_i in *TKList \ ATKList⁻* and adds the corresponding pair $< s_i, H(k_i) >$ in the *List* in M. Finally, it sends the updated message M to its neighbour node.

```
1 begin
2 │   (T, List) ← M ;
3 │   (s₁, h₁), ..., (s_R, h_R) ← List ;
4 │   for hᵢ, (1 ≤ i ≤ R) do
5 │   │   if ∃k ∈ TKList, H(k) = hᵢ then
6 │   │   │   ATKList⁻ = ATKList⁻ ∪ {k} ;
7 │   │   │   REMOVE (sᵢ, hᵢ) from List ;
8 │   │   end
9 │   end
10 │   A' = A − |ATKList⁻| ;
11 │   Select A' twin-keys from TKList \ ATKList⁻: k₁, ..., k_{A'} ;
12 │   Select A' random numbers: s'₁, ..., s'_{A'} ;
13 │   TempATKList = {(s'₁, k₁), ..., (s'_{A'}, k_{A'})} ;
14 │   List = List ∪ {(s'₁, H(k₁)), ..., (s'_{A'}, H(k_{A'}))} ;
15 │   send (T, List) to the next node // (T, List) encrypted with pair-wise key ;
16 end
```

Algorithm 14: FIRSTROUND

Each node, except the CH, executes the procedure SECONDROUND (Algorithm 15) when it receives a liveness announcement message for the second time. The CH executes this procedure when it receives the message for the first time. For each key that the executing node announced in the previous round of the message (line 4), it checks if the corresponding declaration is still in the message M it just received (line 5). If it is so, this means that no other node has removed the declaration from M. The node will then remove the non-alive key from the message. Otherwise, i.e. the key is alive (someone stored the key in its $ATKList^-$), the executing node puts the corresponding key in its list $ATKList^+$. Furthermore, for each declaration in M (line 11), the node checks if it is storing a key (in $TKList$), not yet announced by itself (not in $ATKList^+$), that some other node is asking for liveness (line 12). Note that at this point the node has removed its own declared keys from the $List$ in M: the only keys in M are declared by other nodes, indeed. If the node has such a key k, it adds the key to its $ATKList^-$ (line 13) and removes the corresponding declaration from the $List$ (line 14). The node then checks the number of twin-keys it is using in this aggregation (line 17). If this number is smaller than V, the required number of keys deemed necessary to satisfy the node privacy, it will participate in the aggregation only with the shadow values but not with its own private value (line 19). Otherwise, it will participate also with its private value, and increases the number T of participating nodes (line 21). Finally, the executing node sends the message M to its next neighbour. Note that, when the CH receives the message for the second time, the cluster aggregation process terminates. Then, the tree aggregation is performed as described in point 3.2 of Sect. 6.4.

Figure 6.3 illustrates an example of the result of the procedures shown in this section applied to the same setting used in Fig. 6.2. Keys in bold font are the alive

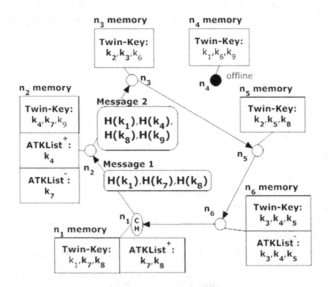

Fig. 6.3 Twin-key liveness announcement example. $A = 3$, $V = 2$

keys while non-bold fonts indicates non-alive key. For example, node CH does not consider alive the key k_1, actually shared with node n_4 that is currently off-line.

6.6.2 Data Aggregation with Shadow Values: Protocol Description

Here we describe the aggregation phase, while each node executes the procedure NODEAGGREGATION (Algorithm 16).In particular, if a node does not have enough alive twin-keys to protect its own private value (line 3), it just does not participate in the aggregation. That is, it does not include its own value in the aggregate (line 3). Otherwise, it initializes the variable x with its own private value d (line 6). Then, for each key in $ATKList^+$ (line 8), it adds $H(Seed, k)$ to x (line 9). Similarly, for each key in $ATKList^-$ (line 11) it removes $H(Seed, k)$ from x (line 12). Finally, the value $x + y$ is sent to the node's next neighbour. Note that $Seed$ is unique for each different data aggregation. For example, it can be a random number broadcast from the BS together with the aggregation request. Also, it could be a time sequence number when the data aggregation is executed, if executed at given interval of time without any request from the BS.

```
 1 begin
 2 │   (T, List) ← M ;
 3 │   (s_1, h_1), ..., (s_R, h_R) ← List ;
 4 │   for (s, k) ∈ TempATKList do
 5 │   │   if ∃(s_i, h_i), s = s_i ∧ H(k) = h_i then
 6 │   │   │   REMOVE (s_i, h_i) from List ;
 7 │   │   else
 8 │   │   │   ATKList^+ = ATKList^+ ∪ {k} ;
 9 │   │   end
10 │   end
11 │   for (s_i, h_i) ∈ List do
12 │   │   if ∃k ∈ TKList \ ATKList^+, H(k) = h_i then
13 │   │   │   ATKList^- = ATKList^- ∪ {k} ;
14 │   │   │   REMOVE (s_i, h_i) from List ;
15 │   │   end
16 │   end
17 │   Num_k = |ATKList^+ ∪ BagToSet(ATKList^-)| ;
18 │   if Num < V then
19 │   │   Avalid = false ;
20 │   else
21 │   │   T = T + 1 ;
22 │   end
23 │   send (T, List) to the next node //encrypted with pair-wise key ;
24 end
```

Algorithm 15: SECONDROUND

```
 1 begin
 2     (y) ← M ;
 3     if Avalid = false then
 4     |   x = 0 ;
 5     else
 6     |   x = d ;
 7     end
 8     for k ∈ ATKList⁺ do
 9     |   x = x + H(Seed, k) ;
10     end
11     for k ∈ ATKList⁻ do
12     |   x = x − H(Seed, k) ;
13     end
14     send (x + y) to the next node //encrypted with pair-wise key ;
15 end
```

Algorithm 16: NODEAGGREGATION

Figure 6.4 illustrates an example of data aggregation on the same setting considered in Fig. 6.2, with the alive twin-keys shown in Fig. 6.3. In this example, the CH node adds its own values d_1, $H(Seed, k_7)$, and $H(Seed, k_8)$ (it has both k_7 and k_8 in its $ATKList^+$). The following node (n_2) adds its own value d_2, adds $H(Seed, k_4)$, and subtracts $H(Seed, k_7)$ (k_4 and k_7 are in $ATKList^+$ and $ATKList^-$ respectively). The resulting message sent by node n_2 has a value that now corresponds to $d_1 + H(Seed, k_4) + d_2 + H(Seed, k_8)$. When the message reaches CH again, it will contain the exact sum of the private values of all the participating nodes.

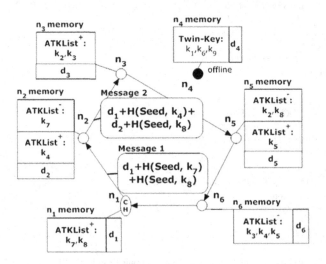

Fig. 6.4 Data aggregation with shadow values. $A = 3, V = 2$

6.6.3 A Complete Protocol Run

Here, we show how to integrate the alive announcement round and the aggregation round together to have a more efficient solution: The two phases can run together using one single message containing both liveness announces and aggregated value. Note that the twin-key liveness announcement requires the message M to be relayed two times along the circuit. Also, note that the second time each node receives the message knows exactly which of the possessed twin-keys it should use in the aggregation. Then, we can think that in the same message relayed in this second round the aggregated value is also routed and updated according to Algorithm 16.

6.7 Security and Complexity Analysis

In this section, we present a thorough analysis of our protocol. In next subsection we study the problem's parameters; in Sect. 6.7.2 we discuss the security features of our proposed protocol; finally, in Sect. 6.7.3, performance analysis is shown. Finally, in Sect. 6.7.4 we compare our protocol with other solutions in literature.

6.7.1 Parameter Study

In this section we provide guidelines for selecting the values to assign to the parameters of our protocol, that is P, K, C, and A. We start reminding that the necessary condition for a node to participate in a cluster-based aggregation is to share A keys with other nodes in the same cluster. Note that, if the above condition does not hold for a node, this node can attempt to join another (neighbouring) cluster it shares enough keys with. In practice, we expect that for a given set of nodes in a cluster there is a reasonably high probability, p_s, that all these nodes share A keys. Set the desired probability p_s, assignment for P, C, A, K satisfying p_s should be found.

For example, we can set $p_s = 0.99$. Then, we can choose the key-pool size $P = 10,000$, the cluster size $C = 20$, and require that each node shares $A = 5$ keys with the other $C - 1$ nodes in the same cluster. Then, the only other parameters we can tune to satisfy p_s is the variable K. To simplify the analysis, without making it less rigorous, we state the following assumptions:

- The K keys assigned to each node are chosen from P with replacement.
- The probability of sharing any of the A keys is independent for the nodes in the cluster. Note that, as observed in [76], this is a feasible approximation.

Using the previous assumptions, the probability for a given node to share at least A keys with the others $C - 1$ nodes is equal to:

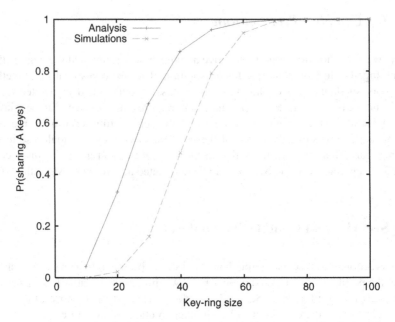

Fig. 6.5 Probability for a node to share at lest A twin-keys in the cluster, varying the key-ring size, K. $P = 10, 000, C = 20, A = 5$

$$p_s = 1 - \sum_{i=0}^{A-1} \binom{K(C-1)}{i} \left(\frac{K}{P}\right)^i \left(\frac{P-K}{P}\right)^{K(C-1)-i} \tag{6.1}$$

In Fig. 6.5 we plot the analytical result (Eq. 6.1) and the simulation result for $P = 10, 000$, $C = 20$ and $A = 5$. From this graph we can observe that $p_s > 0.99$ is achieved for any $K \geq 65$. Note that, once the required p_s is guaranteed, choosing a smaller K increases security. Indeed, an adversary capturing a node will acquire a smaller number of pool's keys.

Then, to satisfy $p_s > 0.99$ for the selected $P = 10, 000$, $C = 20$, and $A = 5$, we chose $K = 65$. This choice of parameters will be used for the following sections.

To show how this parameters choice affects the number of nodes participating in the aggregation phase, in Fig. 6.6 we simulate the protocol, reporting (y-axis) the number of nodes actively participating in the aggregation, while increasing the number of the off-line nodes (x-axis). A non-off-line node actively participates in the aggregation adding its own value and the hashes of its alive twin-keys, if the number of these hashes are at least V. Otherwise, it passively participates adding only the hashes of its alive twin-keys. In Fig. 6.6 we consider different number of agreed twin-keys, $A = 5, 6, 7$. For each of these, we also consider different number of alive twin-keys required for the node active participation ($V = 3, 4, 5$). Note that, for $A = 5$, $V = 5$, the number of off-line nodes significantly affects the active participation of the other nodes. We expect that a similar behaviour could be observed

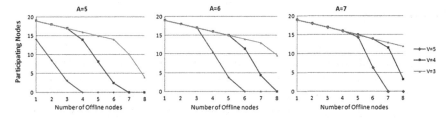

Fig. 6.6 On-line nodes actively participating in the aggregation, while increasing the number of the off-line nodes. $P = 10,000$, $C = 20$, $K = 65$

when $A = V$. However, if we set A to be greater than V, such negative effects are seriously reduced. For example, for $C = 20$, $A = 5$, $V = 3$, if 7 nodes are off-line, on average 10 out of the 13 on-line nodes can actively participate in the aggregation.

6.7.2 Security Analysis

In this section, we analyze the security of our protocol based on the threat model discussed in Sect. 6.3. That is, the aim of the attacker is just to compromise the privacy of the nodes. To reach this goal, We assume that an attacker can: (1) eavesdrop all of the communications in the network, (2) steal information from the BS, and (3) compromise a fraction of the network nodes.

For ease of exposition, we explain the security features of our protocol considering an increasingly powerful attacker. First, we assume that the attacker can just eavesdrop the exchange of messages. Due to the pair-wise encryption between nodes, the attacker cannot obtain any useful information to compromise a single node's privacy.

Then, we consider that the attacker can also steal information from the BS. In this case, we observe that in our protocol, the BS only acts as a receiver of the final aggregation result. There is no other information that can be gathered when compromising the BS. Therefore, the attacker obtains no useful information from the BS to compromise a single node's privacy.

A major threat appears when we assume that the attacker can also capture some nodes. In fact, all of the information stored in the captured nodes become known to the attacker, including pre-distributed keys and the agreed twin-keys. While it is not possible to protect the privacy of the captured nodes, our aim is to protect the privacy of the non-captured nodes. Therefore, we are interested in assessing the probability that the attacker can compromise the privacy of a non-captured node leveraging the information acquired having captured a certain number of nodes.

We assess this probability in two scenarios, hypothesising two different adversary behaviors: the *passive* and *active* one, described below.

6.7.2.1 Passive Attack

In this scenario, we assume that the attacker can only elaborate on the information extracted from the captured nodes (i.e. from the memory) and the received messages.

Each node value is protected by one or more shadow values. The secrecy of the shadow value, in turn, is protected by the secrecy of the twin-keys. To compromise the privacy of non-captured node, n_i, the attacker has to obtain the keys used to generate the shadow values that n_i uses to protect its own privacy. For a node n_i, it will send out a value computed by the following expression (see Algorithm 16):

$$d_i + \sum_{k \in ATKList_i^+} H(k, Seed) - \sum_{k \in ATKList_i^-} H(k, Seed) \qquad (6.2)$$

We refer to this value as the *coated value* of the node n_i. We recall that *Seed* is a one-time-use number broadcasted by the BS together with the aggregation request.

Before starting our analysis we use an example to illustrate how the privacy of non-captured nodes is protected, even after the attacker compromised a significant number of other nodes. Figure 6.7 shows an example with $A = 3$ and $V = 3$. The attacker controls half of the nodes in the cluster: n_1, n_4 and n_5. The following twin-keys are then known to the attacker: k_1, k_2, k_5, k_6, k_7, k_8 (indicated with non-bold font in Fig. 6.7). Besides, k_3 and k_4 remain unknown to the attacker. Also, in the aggregation phase, the attacker is able to obtain the content of the messages received or sent by controlled nodes. In this example, from these messages, the attacker can derive the following equations containing the private values of non-captured nodes:

$$Observed\_v_1 = d_6 - H(K_3, Seed) - H(K_4, Seed) \qquad (6.3)$$

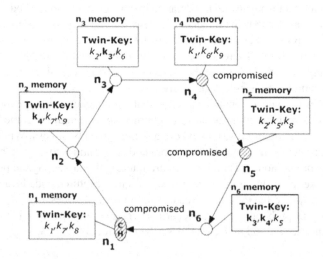

Fig. 6.7 Attack example: Not compromised twin-key. $A = 3$, $V = 3$

$$Observed\_v_2 = d_2 + d_3 + H(K_3, Seed) + H(K_4, Seed) \tag{6.4}$$

In particular, $Observed\_v_1$ is computed as the difference of the content of the messages sent and received by node n_6; $Observed\_v_2$ is computed as the difference between the message sent by n_3 and the one received by n_2. With five unknown variables, this system of equations cannot be solved by the attacker. The privacy of the non-captured nodes: n_2, n_3, n_6, is still protected against the attacker.

In general, the following lemma holds.

Lemma 6.1 *By executing the algorithm described in Sect. 6.6, each private value counted in the aggregation is protected by at least V keys, where $V \leq A$ and A is the number of twin-keys possessed by each node.*

Proof By construction, each node executing the protocol in Sect. 6.6 will participate in the aggregation phase, Algorithm 16, if and only if it finds out at least V alive twin-key it shares with other nodes, Algorithm 15, line 18. □

Next, we give a formal analysis for the probability that the privacy of a non-captured node can be compromised by the attacker. In order to compromise the private value d_i of node n_i, the attacker has to obtain both the coated value and the sum of the shadow values, that is:

$$\sum_{k \in ATKList_i^+} H(k, Seed) - \sum_{k \in ATKList_i^-} H(k, Seed). \tag{6.5}$$

For the coated value we have the following lemma.

Lemma 6.2 *The attacker can compromise the coated value of a node, n_i, if and only if it compromises the two neighbours of n_i in the circuit used during the aggregation (Algorithm 16).*

Proof ($\Rightarrow$) First of all, notice that if the attacker compromises the two neighbours of n_i in the circuit, it can observe the content of the following messages: (i) the aggregation message received by n_i, and (ii) the message sent out from n_i. The difference between the values in these messages corresponds to the coated value of node n_i.

($\Leftarrow$) Assume only one of the neighbours of n_i is compromised. Also, assume the worst scenario where all of the other nodes in the cluster are compromised but node n_i and one of its neighbour, n_{i+1}. Without loss of generality we can assume that n_i appears before n_{i+1} in the circuit. In this case, the attacker would be able to observe: (i) the message received by n_i, and (ii) the messages sent out by n_{i+1}. By the difference of the values in the messages (i) and (ii) the attacker can observe the result of the following expression:

$$d_i + \sum_{k \in ATKList_i^+} H(k, Seed) - \sum_{k \in ATKList_i^-} H(k, Seed)+$$

$$d_{i+1} + \sum_{k \in ATKList_{i+1}^+} H(k, Seed) - \sum_{k \in ATKList_{i+1}^-} H(k, Seed). \qquad (6.6)$$

Once the aggregated value is sent out in the message (ii) by node n_i, neither the value d_i nor d_{i+1} will be removed by other nodes. Hence, the attacker cannot obtain the value of any expression that contains only one of d_i or d_{i+1}. $\square$

Assume the attacker obtained the coated value of a node, n_i, by controlling its two neighbours. Next step is to obtain the n_i's shadow value. There are two kinds of useful information for the attacker to reconstruct the shadow value:

- *Type 1* knowledge. Twin-keys obtained from the captured nodes.
- *Type 2* knowledge. Sum of set of shadow values $H(k, Seed)$, for a key k. The attacker can obtain these values by capturing the neighbours of a node that participate in the aggregation without contributing its own value. Recall that a node, n_i, under the condition that less than V of its twin-keys are alive in the current aggregation phase (Algorithm 15, lines 18–19 and Algorithm 6.6, lines 3–6.), will add to the aggregate just the shadow values computed from its alive twin-keys (without its own value d_i).

For these two types of knowledge the following two observations hold.

Observation 6.1 For a node n_i, *Type 2* knowledge is a subset of *Type 1* knowledge.

On one hand, if the attacker knows all the keys of a node (*Type 1* knowledge), it can compute any possible subset of shadow values, including *Type 2* knowledge. On the other hand, *Type 2* knowledge does not reveal the secret twin-keys.

Observation 6.2 Given the attacker can compromise $w \geq 2$ nodes, its best attack strategy for compromising the privacy of node n_i is the following: (1) By Proposition 6.2, the attacker has to compromise the two neighbours of the target node; (2) The attacker captures the remaining $(w-2)$ nodes selecting every other nodes, in a circuit, following one of the n_i's neighbours.

In fact, by capturing nodes in this way, the attacker will have *Type 1* knowledge over the w captured nodes and will also have the chance to obtain *Type 2* knowledge from the $(w-2)$ nodes between two close captured nodes.

For example, to compromise node n_1 in Fig. 6.8, the best strategy for an attacker that can capture $w = 4$ nodes is: (1) to capture the two neighbours n_2, n_c; (2) to capture the nodes n_4 and n_6 as above described. In this way, the attacker will have *Type 1* knowledge over n_c, n_2, n_4 and n_5. Also, it will have *Type 2* knowledge for n_1, n_3 and n_5.

Let us assume that $w \geq 2$ nodes are compromised using the strategy in Observation 6.2 Based on Observations 6.1 we can consider an upper bound on the attacker's knowledge as has captured also all of the nodes, n_{i+1}, between two consecutive compromised nodes, n_i and n_{i+2}. (except the privacy attack target node). In the above example, when the attacker captures nodes n_2 and n_4, we assume that the node n_3 has also been captured. That is, we bound the *Type 2* knowledge of the attacker by the *Type 1* knowledge.

Fig. 6.8 Best attack strategy

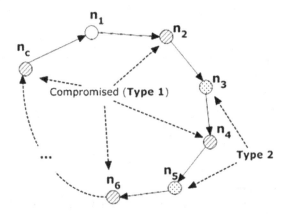

In conclusion, if the attacker controls w nodes, we assume it has *Type 1* knowledge over $2 + 2(w - 2)$ nodes. Referring to the Example in Fig. 6.8, we assume the attacker has *Type 1* knowledge of the nodes n_c, n_2, n_3, n_4, n_5 and n_6.

To ease exposition, in the following analysis, we assume that each twin-key is shared by only two nodes in a cluster. In Fig. 6.9 we report the simulation results for the probability, $p_i(k)$; that is, the probability that the same symmetric key k is considered as twin-key by i nodes in the cluster, assuming the parameters identified in Sect. 6.7.1. From this figure, we can notice that the probability that a key is actually shared between more than two nodes is quite small, i.e. less than 4 % in this example.

Furthermore, we assume that the attacker controls w nodes. From the previous upper bound on the attacker knowledge, the adversary has *Type 1* knowledge for $2 + 2(w - 2) = 2w - 2$ nodes; the probability that the adversary has knowledge of a single key of n_i is $\left(\frac{2w-2}{C-1}\right)$. Then, the upper bound probability that the adversary

Fig. 6.9 Probability, $p_i(k)$, that a twin-key k is shared among i nodes in the cluster

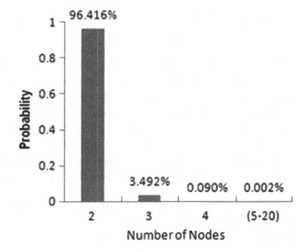

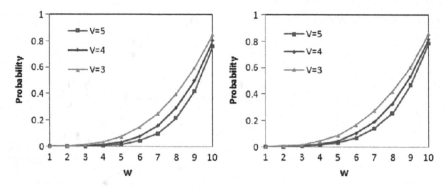

Fig. 6.10 Probability of privacy compromising, varying the number of nodes captured by the attacker. $P = 10,000, C = 20, A = 5$. **a** Analysis result. **b** Simulation result

knows all the V keys a node uses in a data aggregation is:

$$\left(\frac{2w - 2}{C - 1}\right)^V \tag{6.7}$$

Figure 6.10a shows the above probability for the attacker to compromise the privacy of a node by capturing w nodes (x-axis). As an example, for $w = 5$ (the attacker has knowledge of the key ring of $2 + 2(w - 2) = 8$ nodes, each one composed of $K = 65$ symmetric keys) we have that the attack's success probability is some $0.01, 0.03$, and 0.07 for $V = 5, 4$, and 3 respectively. The analytical result shown in Fig. 6.10a is confirmed by the simulation result in Fig. 6.10b. Note that the simulation is done without the assumption of any constraint on the number of nodes sharing a twin-key.

6.7.2.2 Active Attack

In the following we analyze an *active* attacker, that is an attacker that leverages the w controlled nodes to push forward the privacy compromising of a node n_i.

A necessary condition to compromise the privacy of n_i is to control the neighbours of n_i, as previously discussed. Assume the attacker actively controls these nodes during the twin-key liveness announcement phase (described in Sect. 6.6.1). Then, it can be able to enforce n_i not to participate in the aggregation (described Sect. 6.6.2). In fact, using the neighbour of n_i, the attacker can let n_i believe that all of the twin-keys it agreed on in the twin-key agreement (described in Sect. 6.5), are not alive during the current aggregation phase. We observe that this does not imply any privacy violation of node n_i.

However, if the attacker controls the neighbors of n_i during the twin-key agreement (note that the set-up phase is performed only once) it can do something more. Actually, controlling n_i's neighbours the attacker can try to let n_i agree on twin-keys only with the n_i's neighbours (nodes controlled by the attacker). If the attacker does not have enough keys (V), n_i will simply not participate in the aggregation for this cluster. However, if the attacker has enough keys to let n_i agree on V keys known by the attacker, this can pose a serious threat to the protocol: the privacy of the node could be violated. To solve this problem, and also to increase the resilience of our protocol against other kind of attacks, we propose an extension of our protocol by using multiple logical circuits.

6.7.2.3 Using Multiple Logical Circuit to Improve the Protocol Resilience

To address the problem exposed in the previous section, we extend our protocol using multiple logical circuits in the twin-key agreement phase, as shown in Fig. 6.11. Then, when a node initiates a twin-key agreement, it randomly selects one of the available circuits. This requires the attacker to control an higher number of nodes to achieve the same goal as in the single circuit scenario. In general, to compromise the privacy of a given node n_i, the attacker must capture all the other $C - 1$ nodes, if $\frac{C!}{2}$ logical circuit are used.

6.7.2.4 Managing Off-Line Nodes and Message Loss

In Sect. 6.6.1 we described the twin-key liveness announcement mechanism that allows our protocol to be resilient to a node failure. In fact, a correct aggregate can be computed also if some nodes are off-line. For easy of exposition, in Sect. 6.6.1 we described our solution considering only the alive nodes. However, a technique for traversing a circuit with off-line nodes must be used. Here, we briefly explain a simple solution to achieve this goal. We can assume that each node broadcasts the ids of its right and left neighbour, for each logical circuit built. In this way, each node can autonomously know the sequence of nodes for each circuit. Then, after sending a message to its neighbour, the sender node waits for an acknowledge message, *ack*. If the ack is not received, the sender sends the message to the next node in the circuit. For example, in Fig. 6.12 node n_4 sends a message to n_5 and waits for an ack. If the

Fig. 6.11 Using multiple circuits

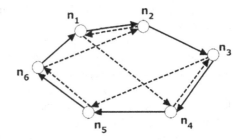

Fig. 6.12 Managing off line nodes

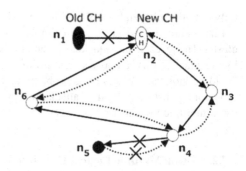

ack is not received within an expected time interval, n_4 sends the message to the next node in the circuit, n_6.

In a similar way, if the CH does not start the protocol within a given time interval, the next node (n_2 in Fig. 6.12) can assume the role of CH sending the first message. Note that message loss problem can be treated in the same way. If the message or the corresponding ack is lost, the destination node is regarded as off line.

6.7.3 Complexity Analysis

In this section, we analyze the complexity of our proposed algorithm from both the communication and the computation point of view.

First, we analyze the complexity of the algorithm in the data aggregation phase (described in Sect. 6.6). Note that this is the major part of the overhead. In fact, such complexity is associated to each single aggregation. Finally, we discuss the complexity of the set-up phase (described in Sect. 6.5) that, instead, is performed only once.

6.7.3.1 Complexity in Data Aggregation Phase

First, we analyze the computation complexity of the data aggregation. We consider the per node worst case overhead. For each agreed twin-key, k, each node has to compute two hash values. One hash is computed for the verification of the liveness announcement of k. The other hash is executed to compute the k's corresponding shadow value added in the aggregated value. Furthermore, each node has to compute two symmetric key encryption and two symmetric key decryption to receive and to send out the required message. Therefore, the computation of each node in the worst case is $2A$ hash computations, 2 symmetric key encryption and 2 symmetric key decryption, considering A agreed twin-keys. The overall computational complexity is in general $O(1)$, with respect to the size of the cluster, C, as reported in Table 6.2

From the communication complexity point of view, each node only needs to receive and to send out two messages. That is, the number of messages is $O(1)$.

Table 6.2 Private data aggregation protocols: Comparing the security features and complexities

	Aggregation type	Encryption type	Privacy versus outsiders	Privacy versus other nodes	Privacy versus BS	Data-loss resilience	Node comput. complexity	Node comm. complexity
Girao et al. CDA protocol [96]	hop-by-hop	CH-to-BS	Yes	No	No	**Yes**	$O(1)$	$O(1)$
Castelluccia et al. protocol [32]	Hop-by-hop	Node-to-BS	Yes	**Yes**	No	No	$O(1)$	$O(1)$
He et al. CPDA protocol [111]	CH	Node-to-CH	Yes	**Yes**	**Yes***	**Yes***	$O(C^2 \log C)$	$O(C)$
He et al. SMART Protocol [111]	Hop-by-hop	Node-to-node	Yes	**Yes**	**Yes**	No	$O(1)$	$O(C)$
Mlaih et al. protocol [154]	Hop-by-hop	Node-to-BS	Yes	**Yes**	No	No	$O(1)$	$O(1)$
Our solution	CH	Node-to-nodes	Yes	**Yes**	**Yes**	**Yes**	$O(1)$	$O(C)$

(* the original protocol can be extended to achieve this property)

However, we should also considered the size of these messages. Each node has to send out the hash values corresponding to its twin-keys that have not yet been declared alive by other nodes (Algorithm 14, lines 10–15). For example, consider the message received by n_2, in Fig. 6.3, during the twin-key liveness announcement. CH sent a message announcing k_1, k_7 and k_8. Now, the set of twin-key for n_2 contains k_4, k_7, and k_9, therefore it will remove k_7 from the message and will add all of its twin-keys not yet declared alive, k_4 and k_9. Consider each node established A twin-keys. In the worst case, $AC/2$ different twin-keys have to be declared in the twin-key liveness announcement message. We recall that we considered a declared (or established) twin-key as the symmetric key shared between a pair of alive nodes. For example, if the same symmetric key, k, is used in two different pairs of nodes it is considered as two different established twin-keys. Then, the worst case message size is $AC/2$. That is, the size of message is $O(C)$. Considering the $O(1)$ number of messages, the overall communication overhead is $O(C)$, as reported in Table 6.2

Next, we estimate the average message size $M_{\text{avg-size}}$. Let L_k be the average number of messages each alive twin-key, k, appears in. Assuming each node is alive, the liveness of all the twin-keys is checked when the message comes back to CH for the first time (when CH receives the liveness message in Algorithm 15 the *List* structure is empty). Each twin-key k, out of the $AC/2$ twin-keys, will be in L_k messages out of the C messages sent in the circuit. Then, we have that:

$$M_{\text{avg-size}} = \frac{\frac{AC}{2}L_k}{C} = \frac{AL_k}{2}. \tag{6.8}$$

Furthermore, given a twin-key k, we define the function $m(k)$ as the number of nodes in the cluster that use the symmetric key k as a twin-key. Then, we have that $L_k = C/m(k)$. We define $p_i(k)$ as the probability that $m(k) = i$, where $i \in [2 \ldots C]$. Equation 6.8 can be rewritten as:

$$M_{\text{avg-size}} = \frac{A}{2} \sum_{i=2}^{C} p_i(k) \frac{C}{i}. \tag{6.9}$$

Finally, rewriting $p_i(k)$ in terms of pool size, P, and key-ring size, K, we have:

$$M_{\text{avg-size}} = \frac{A \sum_{i=2}^{C} \binom{C}{i} \left(\frac{K}{P}\right)^i \left(\frac{P-K}{P}\right)^{c-i} \left(\frac{C}{i}\right)}{2 \sum_{i=2}^{C} \binom{C}{i} \left(\frac{K}{P}\right)^i \left(\frac{P-K}{P}\right)^{c-i}}. \tag{6.10}$$

Assuming each twin-key shared by exactly two nodes, a more tight average message size applies: $AC/4$—that is, $O(C)$ if A is a constant.

6.7.3.2 Complexity in Set-Up Phase

We consider first the communication complexity. Each node has to test, in the worst case, each of its pre-distributed keys to find out the required A twin-keys shared with other nodes. Therefore, up to CK keys will be declared in the message passing through the circuit, where K is the number of the pre-distributed keys in each node and C is the cluster size. Remember that our protocol binds to r the number of keys that a node can declare each time it has the declaration message (Algorithm 10, lines 3–4 and Algorithm 11, lines 37–38). As an upper bound, we can also assume that R is the overall maximum message size. Then, the upper bound for the total number of messages transferred by each node is CK/R.

Finally, we consider the computation complexity under the same assumptions. Each node, in the worst case, has to compute the hash for each of its pre-distributed keys and to encrypt each message it sends out. Altogether, we have K hash computations and CK/R encryptions.

6.7.4 Comparison

In Table 6.2 we summarize the features of our proposal compared with other relevant algorithms present in the literature. The feature *aggregation type* indicates who is responsible for the aggregation: *hop-by-hop* means that each node adds its own value to the aggregate while *CH* means that the local aggregation is performed by the cluster head. The column *encryption type* indicates who are the peers of the encryption considered in the protocol. As an example, *node-to-BS* means that each node encrypts some data that cannot be decrypted until it reaches the BS. Table 6.2 also indicates if the protocol protects privacy against outside eavesdropper, other network nodes or the BS, in columns 3, 4, and 5 respectively. The last two columns denote the per node computational and communication complexity. By *data-loss resilience* we refer whether the BS fails to compute the correct aggregate if a few nodes do not participate in the protocol or if a message is lost. Note that in this chapter, we consider that the aggregation protocol cannot send neither the list of the nodes participating in the current aggregation, nor the list of those who do not participate. We renounced to formulate such an assumption since it can be well the case where such lists are $O(n)$, hence vanishing the savings that an aggregation protocol is supposed to deliver.

6.8 Concluding Remarks

In-network data aggregation is commonly used in sensor network for efficiency reason. However, in privacy sensitive applications, a single node can be interested in contributing its sensed value to compute an aggregate, but it would like to have neither other nodes nor the BS to know the value it contributed. Further, data aggregation has to be resilient to data loss. In this work, we proposed an efficient solution for data aggregation that protect the node privacy, according to the above requirement, against a powerful attacker. In particular, the attacker can eavesdrop the exchanged messages, control a fraction of the network nodes, and also acquire some information from the base station.

Our analysis supports the feasibility of our private aggregation protocol, showing that it is secure, scalable, resilient to data loss, and efficient. To the best of our knowledge, our scheme is the first one that provides the above properties all at once.

Chapter 7
Conclusions and Future Works

The number of areas and problems to which Wireless Sensor Networks are applied continuously grow while known and unknown threats affect this technology. Researchers are called to address the design of efficient protocols that are secure against possible attacks. This book provides several contributions in this direction over the previous state of the art:

- As for the pair-wise key establishment, we designed a new probabilistic solution, the Enhanced Cooperative Channel Establishment (ECCE) Protocol. We compared the performance of ECCE with the most known concurrent schemes via both analysis and simulations. The results showed that the ECCE Protocol presents higher probability for any pair of nodes to establish a secure channel and a higher resilience rate (i.e., the attacker needs a bigger effort to corrupt the channel).
- We proposed a new capture detection approach that leverages the network mobility in order for the nodes to trace the presence of the other nodes. In this framework we further proposed two protocols. The results of an extensive set of simulations show that the newly proposed solutions can be practically implemented in sensor networks and under certain mobility conditions (e.g., a certain average node speed) they perform better than solutions that do not leverage network mobility.
- As for the problem of the clone attack, we observed that the state of the art solution was not practical for WSNs. Then, we proposed a new efficient and distributed protocol for the capture attack detection. We showed that our protocol can be practically implemented in WSN and outperforms the previous state of the art protocol in terms of both efficiency (i.e., energy consumption) and performances (i.e., detection rate).
- We proposed the first solution that can compute the Median of the nodes values and verify if the computed value has been falsified by compromised nodes. Furthermore, we proposed an attack resilient Median computation protocol: The sink node is able not only to verify if the aggregated value has been compromised but also to compute the correct aggregated value (not considering the values of the nodes that do not comply to the protocol).

© Springer Science+Business Media New York 2016
M. Conti, *Secure Wireless Sensor Networks*, Advances in Information Security 65,
DOI 10.1007/978-1-4939-3460-7_7

- Another contribution on data aggregation is related to the privacy of a single node during data aggregation. In this case we proposed the first data aggregation protocol that guarantees the privacy of a node not only against the other nodes but also against the sink node.

  While efforts in securing wireless sensor network have already produced good results, many open problems are still there. We aim to concentrate our future research efforts on the following open research issues:

- The solution for the detection of the node capture can be further improved. The candidate ongoing work is dedicated to investigate how the approach works in other mobility models, rather than only in the considered Random Waypoint Mobility model, and experimentally evaluate the solution on realistic mobility patterns.
- The approach we proposed for the detection of the node capture can be generalized in order to compute other properties that involve all the network. The absence of a node from the network might be seen as a special case of these types global properties.
- In our clone detection solution the presence of a centralized authorities that broadcast a random values is required. We want to extend our solution using distributed mechanisms that allow the nodes to agree an a random value. Furthermore, this mechanism should be resilient to the attacker activities. We observe that such a distributed and efficient mechanism would be useful to all the security protocol that requires the use of a random value shared between the network nodes.
- The clone detection solution proposed is designed for static sensor network. Since it seems to be not trivial to extend the provided solution to a mobile environment we aim to investigate such a possibility.
- We aim to design a reference security architecture for WSNs, with a set of coherent and complete assumptions.
- Another aspect of interest for WSNs is the security and efficiency in actuators-sensors interaction.
- Finally, we aim to investigate specific security problems in underwater WSNs.

References

1. https://standards.ieee.org/about/get/802/802.15.html
2. Agrawal, R., Srikant, R.: Privacy-preserving data mining. In: Proceedings of the 2000 ACM SIGMOD International Conference on Management of data (SIGMOD'00), pp. 439–450 (2000)
3. Akyildiz, I.F., Su, W., Sankarasubramaniam, Y., Cayirci, E.: Wireless sensor networks: a survey. Comput. Netw. (Elsevier) **38**(4), 393–422 (2002)
4. Albers, P., Camp, O., Percher, J., Jouga, B., Mé, L., Puttini., R.: Security in ad hoc networks: a general intrusion detection architecture enhancing trust based approaches. In: Proceedings of the First International Workshop on Wireless Information Systems (WIS'02) (2002)
5. Ambrosin, M., Busold, C., Conti, M., Sadeghi, A.R., Schunter, M.: Updaticator: updating billions of devices by an efficient, scalable and secure software update distribution over untrusted cache-enabled networks. In: Computer Security—ESORICS 2014, vol. 8712, pp. 76–93. Springer International Publishing (2014)
6. Ambrosin, M., Conti, M., Dargahi, T.: On the feasibility of attribute-based encryption on smartphone devices. In: Proceedings of the 2015 Workshop on IoT Challenges in Mobile and Industrial Systems, IoT-Sys'15: MobiSys'15 workshop, pp. 49–54 (2015)
7. Ambrosin, M., Hosseini, H., Mandal, K., Conti, M., Poovendran, R.: Verifiable and privacy-preserving fine-grained data-collection for smart metering. In: Proceedings of the 1st IEEE Workshop on Security and Privacy in Cybermatics (IEEE CNS 2015 workshop: SPiCy 2015), to appear. IEEE (2015)
8. Anderson, R.J.: Security Engineering: A Guide to Building Dependable Distributed Systems. Wiley, New York (2001)
9. Anderson, R.J., Kuhn, M.G.: Tamper resistance—a cautionary note. In: Proceedings of the 2nd USENIX Workshop on Electronic Commerce Proceedings, pp. 1–11 (1996). http://www.cl.cam.ac.uk/mgk25/tamper.html
10. Anderson, R.J., Kuhn, M.G.: Low cost attacks on tamper resistant devices. In: Proceedings of the 5th International Workshop on Security Protocols, pp. 125–136 (1998)
11. Anita, E.M., Geetha, R., Kannan, E.: A novel hybrid key management scheme for establishing secure communication in wireless sensor networks. Wirel. Pers. Commun. **82**(3), 1419–1433 (2015)
12. Ardagna, C., Conti, M., Leone, M., Stefa, J., et al.: An anonymous end-to-end communication protocol for mobile cloud environments. IEEE Trans. Serv. Comput. **7**(3), 373–386 (2014)
13. Aura, T., Nikander, P., Leiwo, J.: Dos-resistant authentication with client puzzles. In: Proceedings of the 8th International Workshop on Security Protocols, pp. 170–177 (2001)

© Springer Science+Business Media New York 2016   157
M. Conti, *Secure Wireless Sensor Networks*, Advances in Information Security 65,
DOI 10.1007/978-1-4939-3460-7

14. Bandyopadhyay, S., Coyle, E.J., Falck, T.: Stochastic properties of mobility models in mobile ad hoc networks. IEEE Trans. Mob. Comput. **6**(11), 1218–1229 (2007)

15. Barbará, D., DuMouchel, W., Faloutsos, C., Haas, P.J., Hellerstein, J.M., Ioannidis, Y.E., Jagadish, H.V., Johnson, T., Ng, R.T., Poosala, V., Ross, K.A., Sevcik, K.C.: The new jersey data reduction report. IEEE Data Eng. Bull. **20**(4), 3–45 (1997)

16. Becher, A., Benenson, Z., Dornseif, M.: Tampering with motes: real-world physical attacks on wireless sensor networks. In: Proceedings of the 3rd International Conference on Security in Pervasive Computing (SPC'06), pp. 104–118 (2006)

17. Ben Jaballah, W., Conti, M., Mosbah, M., Palazzi, C.: Fast and secure multihop broadcast solutions for intervehicular communication. IEEE Trans. Intell. Transp. Syst. **15**(1), 433–450 (2014)

18. Beresford, A.R., Stajano, F.: Location privacy in pervasive computing. IEEE Pervasive Comput. **2**(1), 46–55 (2003). http://dx.doi.org/10.1109/MPRV.2003.1186725

19. Bertossi, A.A., Olariu, S., Pinotti, M.C.: Efficient corona training protocols for sensor networks. Theor. Comput. Sci. **402**(1), 2–15 (2008)

20. Bettstetter, C.: On the minimum node degree and connectivity of a wireless multihop network. In: Proceedings of the 3rd ACM International Symposium on Mobile Ad Hoc Networking and Computing (MobiHoc'02) (2002)

21. Bettstetter, C., Hartmann, C.: Connectivity of wireless multihop networks in a shadow fading environment. In: Proceedings of the 6th ACM International Workshop on Modeling, Analysis and Simulation of Wireless and Mobile Systems (MSWiM'03), pp. 28–32 (2003). http://doi.acm.org/10.1145/940991.940998

22. Blundo, C., De Santis, A., Herzberg, A., Kutten, S., Vaccaro, U., Yung, M.: Perfectly-secure key distribution for dynamic conferences. In: Proceedings of the 12th Annual International Cryptology Conference on Advances in Cryptology (CRYPTO'92), LNCS, vol. 740, pp. 471–486. Springer (1993)

23. Bose, P., Morin, P., Stojmenović, I., Urrutia, J.: Routing with guaranteed delivery in ad hoc wireless networks. Wirel. Netw. **7**(6), 609–616 (2001)

24. Boubiche, D.E., Boubiche, S., Toral-Cruz, H., Pathan, A.S.K., Bilami, A., Athmani, S.: SDAW: secure data aggregation watermarking-based scheme in homogeneous WSNs. Telecommun. Syst. pp. 1–12 (2015)

25. Broch, J., Maltz, D.A., Johnson, D.B., Hu, Y.C., Jetcheva, J.: A performance comparison of multi-hop wireless ad hoc network routing protocols. In: Proceedings of the 4th Annual ACM/IEEE International Conference on Mobile computing and networking (MobiCom'98), pp. 85–97 (1998)

26. Brooks, R., Govindaraju, P., Pirretti, M., Vijaykrishnan, N., Kandemir, M.T.: On the detection of clones in sensor networks using random key predistribution. IEEE Trans. Syst., Man Cybern., Part C: Appl. Rev. **37**(6), 1246–1258 (2007)

27. Brutch, P., Ko, C.: Challenges in intrusion detection for wireless ad-hoc networks. In: Proceedings of the 2003 Symposium on Applications and the Internet Workshops (SAINT'03 Workshops), p. 368. IEEE Computer Society (2003)

28. Burns, B., Brock, O., Levine, B.N.: Mora routing and capacity building in disruption-tolerant networks. Ad Hoc Netw. **6**(4), 600–620 (2008)

29. Buttyán, L., Schaffer, P., Vajda, I.: RANBAR: RANSAC-based resilient aggregation in sensor networks. In: Proceedings of the fourth ACM workshop on Security of Ad Hoc and Sensor Networks (SASN'06), pp. 83–90 (2006)

30. Carman, D., Kruus, P., Matt, B.J.: Constraints and approaches for distributed sensor network security. Technical Report 00-010, NAI Labs (2000)

31. Caruso, A., Urpi, A., Chessa, S., De., S.: GPS-free coordinate assignment and routing in wireless sensor networks. In: 24th Annual Joint Conference of the IEEE Computer and Communications Societies (INFOCOM'05), pp. 150–160 (2005)

32. Castelluccia, C., Mykletun, E., Tsudik, G.: Efficient aggregation of encrypted data in wireless sensor networks. In: The Second Annual International Conference on Mobile and Ubiquitous Systems: Computing, Networking and Services (MobiQuitous'05), pp. 109–117 (2005)

33. Cerpa, A., Elson, J., Estrin, D., Girod, L., Hamilton, M., Zhao, J.: Habitat monitoring: application driver for wireless communications technology. SIGCOMM Comput. Commun. Rev. **31**(2 supplement), 20–41 (2001). http://doi.acm.org/10.1145/844193.844196
34. Chaintreau, A., Hui, P., Diot, C., Gass, R., Scott, J.: Impact of human mobility on opportunistic forwarding algorithms. IEEE Trans. Mob. Comput. **6**(6), 606–620 (2007). http://dx.doi.org/10.1109/TMC.2007.1060
35. Chan, H., Perrig, A.: Security and privacy in sensor networks. IEEE Comput. **36**(10), 103–105 (2003). http://doi.ieeecomputersociety.org/10.1109/MC.2003.1236475
36. Chan, H., Perrig, A.: PIKE: Peer intermediaries for key establishment in sensor networks. In: 24th Annual Joint Conference of the IEEE Computer and Communications Societies (INFOCOM'05), pp. 524–535 (2005)
37. Chan, H., Perrig, A., Song, D.: Random key predistribution schemes for sensor networks. In: Proceedings of the 2003 IEEE Symposium on Security and Privacy (S&P'03), pp. 197–213 (2003)
38. Chan, H., Perrig, A., Song, D.: Secure hierarchical in-network aggregation in sensor networks. In: Proceedings of the 13th ACM Conference on Computer and Communications Security (CCS'06), pp. 278–287 (2006)
39. Chao, L., Tsui, C.Y., Ki, W.H.: Vibration energy scavenging and management for ultra low power applications. In: Proceedings of the 2007 International Symposium on Low Power Electronics and Design (ISLPED'07), pp. 316–321 (2007). http://doi.acm.org/10.1145/1283780.1283848
40. Chen, G., Branch, J.W., Szymanski, B.K.: Local leader election, signal strength aware flooding, and routeless routing. In: Proceedings of the 19th International Parallel and Distributed Processing Symposium (IPDPS'05) (2005)
41. Choi, H., Zhu, S., La Porta, T.F.: SET: Detecting node clones in sensor networks. In: Proceedings of IEEE 3rd International Conference on Security and Privacy in Communication Networks (SecureComm'07) (2007)
42. Cocks, C.: An identity based encryption scheme based on quadratic residues. In: Proceedings of the 8th IMA International Conference on Cryptography and Coding, pp. 360–363 (2001)
43. Considine, J., Li, F., Kollios, G., Byers, J.: Approximate aggregation techniques for sensor databases. In: Proceedings of the 20th International Conference on Data Engineering (ICDE'04), pp. 449–460 (2004)
44. Conti, M., Crispo, B., Fernandes, E., Zhauniarovich, Y.: Crêpe: a system for enforcing fine-grained context-related policies on android. IEEE Trans. Inf. Forensics Secur. **7**(5), 1426–1438 (2012)
45. Conti, M., Di Pietro, R., Gabrielli, A., Mancini, L.V., Mei, A.: The smallville effect: social ties make mobile networks more secure against node capture attack. In: Proceedings of the 8th ACM International Workshop on Mobility Management and Wireless Access, pp. 99–106 (2010)
46. Conti, M., Di Pietro, R., Mancini, L.V.: Secure cooperative channel establishment in wireless sensor networks. In: Proceedings of the Fourth Annual IEEE International Conference on Pervasive Computing and Communications Workshops (PERCOMW'06), pp. 327–331 (2006)
47. Conti, M., Di Pietro, R., Mancini, L.V.: ECCE: Enhanced Cooperative Channel Establishment for secure pair-wise communication in wireless sensor networks. Ad Hoc Netw. (Elsevier) **5**(1), 49–62 (2007)
48. Conti, M., Di Pietro, R., Mancini, L.V., Mei, A.: A randomized, efficient, and distributed protocol for the detection of node replication attacks in wireless sensor networks. In: Proceedings of the Eighth ACM International Symposium on Mobile Ad Hoc Networking and Computing (MobiHoc'07), pp. 80–89 (2007)
49. Conti, M., Di Pietro, R., Mancini, L.V., Mei, A.: Mobility and cooperation to thwart node capture attacks in manets. EURASIP J. Wirel. Commun. Netw. **2009**, 1–13 (2009)
50. Conti, M., Di Pietro, R., Spognardi, A.: Clone wars: distributed detection of clone attacks in mobile wsns. J. Comput. Syst. Sci. **80**(3), 654–669 (2014)

51. Conti, M., Dragoni, N., Gottardo, S.: MITHYS: Mind the hand you shake-protecting mobile devices from SSL usage vulnerabilities. In: Security and Trust Management, pp. 65–81. Springer, Berlin (2013)
52. Conti, M., Pietro, R.D., Mancini, L.V., Mei, A.: Requirements and open issues in distributed detection of node identity replicas in WSN. In: Proceedings of the 2006 IEEE International Conference on Systems, Man, and Cybernetics (SMC'06), Special Session on Wireless Sensor Networks, pp. 1468–1473 (2006)
53. Conti, M., Pietro, R.D., Mancini, L.V., Mei, A.: Emergent properties: detection of the node-capture attack in mobile wireless sensor networks. In: Proceedings of the First ACM Conference on Wireless Network Security (WiSec'08), pp. 214–219 (2008)
54. Conti, M., Pietro, R.D., Mancini, L.V., Mei, A.: Distributed data source verification in wireless sensor networks. Inf. Fusion 10(4), 342–353 (2009)
55. Conti, M., Pietro, R.D., Mancini, L.V., Mei, A.: Distributed detection of clone attacks in wireless sensor networks. IEEE Trans. Dependable Secur. Comput. 8(5), 685–698 (2011)
56. Conti, M., Pietro, R.D., Mancini, L.V., Spognardi, A.: RIPP-FS: an RFID identification, privacy preserving protocol with forward secrecy. In: Proceedings of the Fifth Annual IEEE International Conference on Pervasive Computing and Communications Workshops (PERCOMW'07), pp. 229–234 (2007)
57. Conti, M., Pietro, R.D., Mancini, L.V., Spognardi, A.: Thwarting de-synchronization attack for a class of rfid security protocols. Technical Report TR-06-08, Dipartimento di Informatica, Università di Roma "La Sapienza" (2008)
58. Conti, M., Willemsen, J., Crispo, B.: Providing source location privacy in wireless sensor networks: a survey. IEEE Commun. Surv. Tutor. 15(3), 1238–1280 (2013)
59. Conti, M., Zachia-Zlatea, I., Crispo, B.: Mind how you answer me!: transparently authenticating the user of a smartphone when answering or placing a call. In: Proceedings of the 6th ACM Symposium on Information, Computer and Communication Security, pp. 249–259. ACM (2011)
60. Conti, M., Zhang, L., Roy, S., Di Pietro, R., Jajodia, S., Mancini, L.V.: Privacy-preserving robust data aggregation in wireless sensor networks. Secur. Commun. Netw. 2(2), 195–213 (2009)
61. Cormode, G., Muthukrishnan, S.: An improved data stream summary: the count-min sketch and its applications. In: Proceedings of Latin American Theoretical Informatics (LATIN'04), pp. 29–38 (2004)
62. Coyle, G., Boydell, L., Brown, L.: Home telecare for the elderly. J. Telemed. Telecare 1, 183–184 (1995)
63. Cramer, R., Damgard, I., Dziembowski, S.: On the complexity of verifiable secret sharing and multiparty computation. In: Proceedings of the Thirty-Second Annual ACM Symposium on Theory of Computing (STOC'00), pp. 325–334 (2000)
64. Curtmola, R., Kamara, S.: A mechanism for communication-efficient broadcast encryption over wireless ad hoc networks. Electron. Note Theor. Comput. Sci. (Elsevier) 171(1), 57–69 (2007)
65. Daly, E.M., Haahr, M.: Social network analysis for routing in disconnected delay-tolerant manets. In: Proceedings of the 8th ACM International Symposium on Mobile Ad Hoc Networking and Computing (MobiHoc'07), pp. 32–40 (2007)
66. Dargahi, T., Javadi, H.H., Hosseinzadeh, M.: Application-specific hybrid symmetric design of key pre-distribution for wireless sensor networks. Secur. Commun. Netw. 8(8), 1561–1574 (2015)
67. Deb, B., Bhatnagar, S., Nath, B.: Reinform: reliable information forwarding using multiple paths in sensor networks. In: Proceedings of the 28th Annual IEEE International Conference on Local Computer Networks (LCN'03), p. 406 (2003)
68. Demirbas, M., Song, Y.: An RSSI-based scheme for sybil attack detection in wireless sensor networks. In: Proceedings of the 2006 International Symposium on World of Wireless, Mobile and Multimedia Networks (WOWMOM'06), pp. 564–570 (2006)

69. Deng, J., Han, R., Mishra, S.: Countermeasures against traffic analysis attacks in wireless sensor networks. In: Proceedings of the First International Conference on Security and Privacy for Emerging Areas in Communications Networks (SecureComm'05), pp. 113–126 (2005). http://dx.doi.org/10.1109/SECURECOMM.2005.16

70. Deng, J., Han, R., Shivakant, M.: INSENS: INtrusion-tolerent routing in wireless SEnsor Networks. Technical Report CU-CS-939-02, Department of Computer Science, University of Colorado (2002)

71. Derhab, A., Badache, N.: A self-stabilizing leader election algorithm in highly dynamic ad hoc mobile networks. IEEE Trans. Parallel Distrib. Syst. **19**(7), 926–939 (2008)

72. Di Pietro, R., Mancini, L.V.: Intrusion Detection Systems, Advances in Information Security, vol. 38, Springer, New York (2008)

73. Di Pietro, R., Mancini, L.V., Jajodia, S.: Providing secrecy in key management protocols for large wireless sensors networks. J. Ad Hoc Netw. (Elsevier) **1**(4), 455–468 (2003)

74. Di Pietro, R., Mancini, L.V., Mei, A.: Efficient and resilient key discovery based on pseudo-random key pre-deployment. In: Proceedings of the 18th International Parallel and Distributed Processing Symposium (IPDPS'04), pp. 217–224 (2004)

75. Di Pietro, R., Mancini, L.V., Mei, A.: Energy efficient node-to-node authentication and communication confidentiality in wireless sensor networks. Wirel. Netw. (Kluwer Academic) **12**(6), 709–721 (2006)

76. Di Pietro, R., Mancini, L.V., Mei, A., Panconesi, A., Radhakrishnan, J.: Redoubtable sensor networks. ACM Trans. Inf. Syst. Secur. **11**(3), 1–22 (2008)

77. Diffie, W., Hellman, M.E.: New directions in cryptography. IEEE Trans. Inf. Theory **IT-22**(6), 644–654 (1976). http://citeseer.ist.psu.edu/diffie76new.html

78. Domingo-Ferrer, J.: A provably secure additive and multiplicative privacy homomorphism. In: Proceedings of the 5th International Conference on Information Security (ISC'02), pp. 471–483 (2002)

79. Douceur, J.R.: The sybil attack. In: Proceedings of the 1st International Workshop on Peer-to-Peer Systems (IPTPS'02), pp. 251–260 (2002)

80. Du, W., Deng, J., Han, Y.S., Varshney, P.K.: A pairwise key pre-distribution scheme for wireless sensor networks. In: Proceedings of the 10th ACM Conference on Computer and Communications Security (CCS'03), pp. 42–51 (2003). http://doi.acm.org/10.1145/948109.948118

81. Dubhashi, D., Häggström, O., Orecchia, L., Panconesi, A., Petrioli, C., Vitaletti, A.: Localized techniques for broadcasting in wireless sensor networks. Algorithmica **49**(4), 412–446 (2007). http://dx.doi.org/10.1007/s00453-007-9092-8

82. Duri, S., Gruteser, M., Liu, X., Moskowitz, P., Perez, R., Singh, M., Tang, J.M.: Framework for security and privacy in automotive telematics. In: Proceedings of the 2nd International Workshop on Mobile Commerce (WMC'02), pp. 25–32 (2002)

83. Elson, J., Estrin, D.: Time synchronization for wireless sensor networks. In: Proceedings of the 15th International Parallel and Distributed Processing Symposium (IPDPS'01), pp. 1965–1970 (2001)

84. Elson, J., Girod, L., Estrin, D.: Fine-grained network time synchronization using reference broadcasts. SIGOPS Oper. Syst. Rev. **36**(SI), 147–163 (2002)

85. Eschenauer, L., Gligor, V.D.: A key-management scheme for distributed sensor networks. In: Proceedings of the 9th ACM Conference on Computer and Communications Security (CCS'02), pp. 41–47 (2002)

86. Estrin, D., Govindan, R., Heidemann, J., Kumar, S.: Next century challenges: scalable coordination in sensor networks. In: Proceedings of the 5th Annual ACM/IEEE International Conference on Mobile Computing and Networking (MobiCom'99), pp. 263–270 (1999)

87. Evfimievski, A., Srikant, R., Agrawal, R., Gehrke, J.: Privacy preserving mining of association rules. In: Proceedings of 8th ACM SIGKDD International Conference on Knowledge Discovery and Data Mining (KDD'02) (2002)

88. Fernandes, E., Crispo, B., Conti, M.: FM 99.9, radio virus: exploiting fm radio broadcasts for malware deployment. IEEE Trans. Inf. Forensics Secur. **8**(6), 1027–1037 (2013)

89. Frikken, K.: An efficient integrity-preserving scheme for hierarchical sensor aggregation. In: Proceedings of the First ACM Conference on Wireless Network Security (WiSec'08), pp. 68–76 (2008)

90. Fu, F., Liu, J., Yin, X.: Space-time related pairwise key predistribution scheme for wireless sensor networks. In: Proceedings of the 2007 International Conference on Wireless Communications, Networking and Mobile Computing (WiCom 2007), pp. 2692–2696 (2007)

91. Fung, W.F., Sun, D., Gehrke, J.: Cougar: the network is the database. In: Proceedings of the 2002 ACM SIGMOD International Conference on Management of Data (SIGMOD'02), pp. 621–621 (2002)

92. Ganeriwal, S., Pöpper, C., Čapkun, S., Srivastava, M.B.: Secure time synchronization in sensor networks. ACM Trans. Inf. Syst. Secur. 11(4), 1–35 (2008)

93. Ganeriwal, S., Sribastava, M.B.: Reputation-based framework for highly integrity sensor networks. In: Proceedings of ACM Workshop on Security of Sensor and Adhoc Networks (SASN'04) (2004)

94. Ganesan, D., Govindan, R., Shenker, S., Estrin, D.: Highly-resilient, energy-efficient multipath routing in wireless sensor networks. SIGMOBILE Mob. Comput. Commun. Rev. 5(4), 11–25 (2001). http://doi.acm.org/10.1145/509506.509514

95. Gaubatz, G., Kaps, J.P., Sunar, B.: Public key cryptography in sensor networks—revisited. In: Security in Ad-hoc and Sensor Networks, pp. 2–18. Springer (2005)

96. Girao, J., Westhoff, D., Schneider, M.: CDA: concealed data aggregation for reverse multicast traffic in wireless sensor networks. In: 2005 IEEE International Conference on Communications (ICC 2005), pp. 3044–3049 (2005)

97. Giuffrida, C., Majdanik, K., Conti, M., Bos, H.: I sensed it was you: authenticating mobile users with sensor-enhanced keystroke dynamics. In: Detection of Intrusions and Malware, and Vulnerability Assessment, pp. 92–111. Springer, New York (2014)

98. Gligor, V.D.: Emergent properties in ad-hoc networks: a security perspective. In: Proceedings of the 4th ACM Workshop on Wireless Security (WiSe'05), p. 55 (2005)

99. Goldreich, O.: Foundations of Cryptography: Basic Tools. Cambridge University Press, Cambridge (2001). http://www.wisdom.weizmann.ac.il/oded/foc-book.html

100. Greenwald, M., Khanna, S.: Space-efficient online computation of quantile summaries. In: Proceedings of the 2001 ACM SIGMOD International Conference on Management of Data (SIGMOD'01), pp. 58–66 (2001)

101. Greenwald, M.B., Khanna, S.: Power-conserving computation of order-statistics over sensor networks. In: Proceedings of the 23rd ACM SIGMOD-SIGACT-SIGART Symposium on Principles of Database Systems (PODS'04), pp. 275–285 (2004)

102. Grossglauser, M., Vetterli, M.: Locating nodes with EASE: last encounter routing in ad hoc networks through mobility diffusion. In: Proceedings of the 22nd Annual Joint Conference of the IEEE Computer and Communications Societies (INFOCOM'03), pp. 1954–1964 (2003)

103. Gruteser, M., Grunwald, D.: Anonymous usage of location-based services through spatial and temporal cloaking. In: Proceedings of the 1st International Conference on Mobile Systems, Applications and Services (MobiSys'03), pp. 31–42 (2003)

104. Gruteser, M., Grunwald, D.: A methodological assessment of location privacy risks in wireless hotspot networks. In: Proceedings of the First International Conference on Security in Pervasive Computing, pp. 10–24 (2004)

105. Gruteser, M., Schelle, G., Jain, A., Han, R., Grunwald, D.: Privacy-aware location sensor networks. In: Proceedings of the 9th Conference on Hot Topics in Operating Systems (HOTOS'03), pp. 28–28 (2003)

106. Gu, W., Wang, X., Chellappan, S., Xuan, D., Lai, T.: Defending against search-based physical attacks in sensor networks. In: IEEE International Conference on Mobile Adhoc and Sensor Systems Conference (MASS'05), p. 527 (2005)

107. Gura, N., Patel, A., Wander, A., Eberle, H., Shantz, S.C.: Comparing elliptic curve cryptography and RSA on 8-bit CPUs. Cryptographic Hardware and Embedded Systems pp. 119–132 (2004)

108. Halpern, J., Teague, V.: Rational secret sharing and multiparty computation: extended abstract. In: Proceedings of the Thirty-Sixth Annual ACM Symposium on Theory of Computing (STOC'04), pp. 623–632 (2004)
109. Hartung, C., Balasalle, J., Han, R.: Node compromise in sensor networks: the need for secure systems. Technical Report CUCS-990-05 (2005)
110. Hayashibara, N., Cherif, A., Katayama, T.: Failure detectors for large-scale distributed systems. In: Proceedings of the 21st IEEE Symposium on Reliable Distributed Systems (SRDS'02), pp. 404–409 (2002)
111. He, W., Liu, X., Nguyen, H., Nahrstedt, K., Abdelzaher, T.: Pda: Privacy-preserving data aggregation in wireless sensor networks. In: 26th Annual IEEE Conference on Computer Communications (INFOCOM 2007), pp. 2045–2053 (2007)
112. Hengartner, U., Steenkiste, P.: Protecting access to people location information. In: Proceedings of the First International Conference on Security in Pervasive Computing (SPC'03), pp. 25–38 (2003)
113. Hill, J., Szewczyk, R., Woo, A., Hollar, S., Culler, D., Pister, K.: System architecture directions for networked sensors. ACM SIGOPS Oper. Syst. Rev. **34**(5), 93–104 (2000)
114. Hsin, C., Liu, M.: A distributed monitoring mechanism for wireless sensor networks. In: Proceedings of the 1st ACM Workshop on Wireless Security (WiSe'02), pp. 57–66 (2002)
115. Hsin, C., Liu, M.: Self-monitoring of wireless sensor networks. Comput. Commun. (Elsevier) **29**(4), 462–476 (2006)
116. Microsoft: Microsoft Windows CE. http://www.microsoft.com/windows/embedded/ce/ (2008)
117. PalmOS: The PalmOS Platform. http://www.palmos.com/platform/architecture.html (2008)
118. eCos: The eCos Operating System. http://www.redhat.com/ecos (2008)
119. Crossbow Technology Inc. http://www.xbow.com (2008)
120. ZigBee Working Group. http://www.zigbee.org (2008)
121. Hu, Y.C., Perrig, A., Johnson, D.B.: Packet leashes: a defense against wormhole attacks in wireless networks. In: Proceedings of the 22nd Annual Joint Conference of the IEEE Computer and Communications Societies (INFOCOM'03), pp. 1976–1986 (2003)
122. Hwang, J., Kim, Y.: Revisiting random key pre-distribution schemes for wireless sensor networks. In: Proceedings of the 2nd ACM Workshop on Security of Ad Hoc and Sensor Networks (SASN'04), pp. 43–52 (2004)
123. Hyytiä, E., Lassila, P., Virtamo, J.: Spatial node distribution of the random waypoint mobility model with applications. IEEE Trans. Mob. Comput. **5**(6), 680–694 (2006)
124. Information Processing Technology Office (IPTO) Defense Advanced Research Projects Agency (DARPA): BAA 07-46 LANdroids Broad Agency Announcement (2007)
125. Intanagonwiwat, C., Govindan, R., Deborah, E.: Directed diffusion: a scalable and robust communication paradigm for sensor networks. In: Proceedings of the 6th Annual International Conference on Mobile Computing and Networking (MobiCom'00), pp. 56–67 (2000)
126. Jain, R., Chlamtac, I.: The P^2 algorithm for dynamic calculation of quantiles and histograms without storing observations. Commun. ACM **28**(10), 1076–1085 (1985)
127. Kahn, J.M., Katz, R.H., Pister, K.J.: Next century challenges: mobile networking for "smart dust". In: Proceedings of the 5th Annual ACM/IEEE International Conference on Mobile Computing and Networking (MobiCom'99), pp. 271–278 (1999)
128. Karlof, C., Wagner, D.: Secure routing in wireless sensor networks: attacks and countermeasures. In: Proceedings of the First IEEE International Workshop on Sensor Network Protocols and Applications, pp. 113–127 (2003)
129. Karp, B., Kung, H.T.: GPSR: Greedy perimeter stateless routing for wireless networks. In: Proceedings of the 6th Annual ACM/IEEE International Conference on Mobile Computing and Networking (MobiCom'00), pp. 243–254 (2000)
130. Kasten, O., Langheinrich, M.: First experiences with bluetooth in the smart-its distributed sensor network. In: Workshop on Ubiquitous Computing and Communications, PACT (2001)
131. Kaya, T., Lin, G., Noubir, G., Yilmaz, A.: Secure multicast groups on ad hoc networks. In: Proceedings of the 1st ACM Workshop on Security of Ad Hoc and Sensor Networks (SASN'03), pp. 94–102 (2003)

132. Kömmerling, O., Kuhn, M.G.: Design principles for tamper-resistant smartcard processors. In: Proceedings of the USENIX Workshop on Smartcard Technology on USENIX Workshop on Smartcard Technology (WOST'99), pp. 2–2 (1999)

133. Kong, J., Luo, H., Xu, K., Gu, D.L., Gerla, M., Lu, S.: Adaptive Security for Multi-layer Ad Hoc Networks. Special Issue of Wireless Communications and Mobile Computing. Wiley Interscience Press, New York (2002)

134. Kumar, V., Madria, S.: Pip: Privacy and integrity preserving data aggregation in wireless sensor networks. In: Proceeding of the 32nd International Symposium on Reliable Distributed Systems (SRDS'13), pp. 10–19. IEEE (2013)

135. Kwon, S., Shroff, N.B.: Paradox of shortest path routing for large multi-hop wireless networks. In: Proceeding of the 24th Annual Joint Conference of the IEEE Computer and Communications Societies (INFOCOM'07), pp. 1001–1009 (2007)

136. Law, Y.W., Doumen, J., Hartel, P.: Survey and benchmark of block ciphers for wireless sensor networks. ACM Trans. Sens. Netw. 2(1), 65–93 (2006). http://doi.acm.org/10.1145/1138127.1138130

137. Lazos, L., Poovendran, R.: Secure broadcast in energy-aware wireless sensor networks. In: Proceedings of IEEE International Symposium on Advances in Wireless Communications (ISWC'02) (2002)

138. Lazos, L., Poovendran, R.: Energy-aware secure multicast communication in ad-hoc networks using geographic location information. In: Proceedings of the 2003 IEEE International Conference on Acoustics, Speech, and Signal Processing (ICASSP'03) 4 (2003)

139. Lazos, L., Poovendran, R.: Serloc: robust localization for wireless sensor networks. ACM Trans. Sens. Netw. 1(1), 73–100 (2005). http://doi.acm.org/10.1145/1077391.1077395

140. Liang, Z., Shi, W.: Enforcing cooperative resource sharing in untrusted P2P computing environments. Mob. Netw. Appl. 10(6), 971–983 (2005). http://doi.acm.org/10.1145/1160125.1160140

141. Liang, Z., Shi, W.: Pet: A personalized trust model with reputation and risk evaluation for P2P resource sharing. In: Proceedings of the 38th Annual Hawaii International Conference on System Sciences (HICSS'05), pp. 201–202 (2005)

142. Lingxuan Hu, D.E.: Using directional antennas to prevent wormhole attacks. In: Proceedings of the 11th Annual Network and Distributed System Security Symposium (NDSS'03) (2003)

143. Liu, D., Ning, P.: Efficient distribution of key chain commitments for broadcast authentication in distributed sensor networks. In: Proceedings of the 10th Network and Distributed System Security Symposium (NDSS'03), pp. 263–276 (2003)

144. Liu, D., Ning, P.: Establishing pairwise keys in distributed sensor networks. In: Proceedings of the 10th ACM Conference on Computer and Communications Security (CCS'03), pp. 52–61 (2003)

145. Liu, D., Ning, P.: Multilevel μtesla: broadcast authentication for distributed sensor networks. ACM Trans. Embed. Comput. Syst. 3(4), 800–836 (2004). http://doi.acm.org/10.1145/1027794.1027800

146. Liu, H., Wan, P.J., Liu, X., Yao, F.: A distributed and efficient flooding scheme using 1-hop information in mobile ad hoc networks. IEEE Trans. Parallel Distrib. Syst. 18(5), 658–671 (2007)

147. Luo, J., Hubaux, J.P.: Joint mobility and routing for lifetime elongation in wireless sensor networks. In: 24th Annual Joint Conference of the IEEE Computer and Communications Societies (INFOCOM'05) (2005)

148. Madden, S., Franklin, M.J., Hellerstein, J.M., Hong, W.: TAG: a tiny aggregation service for ad-hoc sensor networks. In: Proceedings of the 5th Symposium on Operating Systems Design and Implementation (OSDI'02), pp. 131–146 (2002)

149. Malan, D.J., Welsh, M., Smith., M.D.: A public-key infrastructure for key distribution in tinyos based on elliptic curve cryptography. In: Proceedings of the 1st IEEE International Conference on Sensor and Ad Hoc Communications and Networks (SECON'04), pp. 71–80 (2005)

150. Manku, G.S., Rajagopalan, S., Lindsay, B.G.: Approximate medians and other quantiles in one pass and with limited memory. SIGMOD Rec. **27**(2), 426–435 (1998)
151. Marconi, L., Di Pietro, R., Crispo, B., Conti, M.: Time warp: how time affects privacy in lbss. In: Proceedings of the Information and Communications Security, pp. 325–339. Springer, Berlin (2010)
152. Mei, A., Stefa, J.: Routing in outer space: fair traffic load in multi-hop wireless networks. In: Proceedings of the ACM International Symposium on Mobile Ad Hoc Networking and Computing (MobiHoc'08) (2008)
153. Merkle, R.C.: A digital signature based on a conventional encryption function. In: Proceeding of the Conference on the Theory and Applications of Cryptographic Techniques on Advances in Cryptology (CRYPTO'87), pp. 369–378 (1988)
154. Mlaih, E., Aly, S.: Secure hop-by-hop aggregation of end-to-end concealed data in wireless sensor networks. In: Proceedings of the 27th IEEE Conference on Computer Communications (INFOCOM 2008) pp. 1–6 (2008)
155. Molnar, D., Wagner, D.: Privacy and security in library RFID: issues, practices, and architectures. In: Proceedings of the 11th ACM Conference on Computer and Communications Security (CCS'04), pp. 210–219 (2004)
156. Munro, J.I., Paterson, M.S.: Selection and sorting with limited storage. Theor. Comput. Sci. **12**, 315–323 (1980)
157. Myles, G., Friday, A., Davies, N.: Preserving privacy in environments with location-based applications. IEEE Pervasive Comput. **2**(1), 56–64 (2003). http://dx.doi.org/10.1109/MPRV.2003.1186726
158. Nath, S., Gibbons, P.B., Seshan, S., Anderson, Z.R.: Synopsis diffusion for robust aggregation in sensor networks. In: Proceedings of the 2nd International Conference on Embedded Networked Sensor Systems (SenSys'04), pp. 250–262 (2004)
159. Newsome, J., Shi, E., Song, D., Perrig, A.: The sybil attack in sensor networks: analysis and defenses. In: Proceedings of the 3rd ACM International Symposium on Information Processing in Sensor Networks (IPSN'04), pp. 259–268 (2004)
160. Newsome, J., Song, D.X.: Gem: graph embedding for routing and data-centric storage in sensor networks without geographic information. In: Proceedings of the 1st International Conference on Embedded Networked Sensor Systems (SenSys'03), pp. 76–88 (2003)
161. Okazaki, Y., Sato, I., Goto, S.: A new intrusion detection method based on process profiling. In: Proceedings of the 2002 Symposium on Applications and the Internet (SAINT'02), pp. 82–91 (2002)
162. Oram, A. (ed.): Peer-to-Peer: Harnessing the Power of Disruptive Technologies. O'Reilly and Associates Inc., Sebastopol (2001)
163. Orecchia, L., Panconesi, A., Petrioli, C., Vitaletti, A.: Localized techniques for broadcasting in wireless sensor networks. In: Proceedings of the 2004 ACM Joint Workshop on Foundations of Mobile Computing (DIALM-POMC'04) (2004)
164. Ortolani, S., Conti, M., Crispo, B., Pietro, R.D.: Events privacy in WSNs: A new model and its application. In: Proceedings of the 2011 IEEE International Symposium on a World of Wireless, Mobile and Multimedia Networks (WoWMoM), pp. 1–9. IEEE (2011)
165. Ozdemir, S., Peng, M., Xiao, Y.: Prda: polynomial regression-based privacy-preserving data aggregation for wireless sensor networks. Wirel. Commun. Mob. Comput. **15**(4), 615–628 (2015)
166. Ozturk, C., Zhang, Y., Trappe, W.: Source-location privacy in energy-constrained sensor network routing. In: Proceedings of the 2nd ACM Workshop on Security of Ad Hoc and Sensor Networks (SASN'04), pp. 88–93 (2004)
167. Papadimitratos, P., Haas, Z.: Secure routing for mobile ad hoc networks. In: Proceedings of the SCS Communication Networks and Distributed Systems Modeling and Simulation Conference, pp. 27–31 (2002)
168. Papadimitratos, P., Poturalski, M., Schaller, P.D., Basin, P.L., Čapkun, S., Hubaux, J.P.: Secure neighborhood discovery: a fundamental element for mobile ad hoc networking. IEEE Commun. Mag. **46**(2), 132–139 (2008)

169. Parno, B., Perrig, A., Gligor, V.D.: Distributed detection of node replication attacks in sensor networks. In: Proceedings of the 2005 IEEE Symposium on Security and Privacy (S&P'05), pp. 49–63 (2005)
170. Patt-Shamir, B.: A note on efficient aggregate queries in sensor networks. Theor. Comput. Sci. **370**(1–3), 254–264 (2007)
171. Perrig, A., Stankovic, J., Wagner, D.: Security in wireless sensor networks. Commun. ACM **47**(6), 53–57 (2004). http://doi.acm.org/10.1145/990680.990707
172. Perrig, A., Szewczyk, R., Wen, V., Culler, D., Tygar, J.D.: SPINS: security protocols for sensor networks. Wirel. Netw. **8**(5), 521–534 (2002)
173. Perrig, A., Szewczyk, R., Wen, V., Culler, D.E., Tygar, J.D.: SPINS: security protocols for sensor networks. In: Proceedings of the 7th Annual ACM/IEEE International Conference on Mobile Computing and Networking (MobiCom'01), pp. 189–199 (2001)
174. Pietro, R.D.: Security issues for wireless sensor networks. PhD in computer science, Dipartimento di Informatica—Università degli Studi di Roma "La Sapienza" (2004)
175. Pietro, R.D., Mancini, L.V.: Security and privacy issues of handheld and wearable wireless devices. Commun. ACM **46**(9), 74–79 (2003)
176. Pietro, R.D., Mancini, L.V., Law, Y., Etalle, S., Havinga, P.: LKHW: a directed diffusion-based secure multicast scheme for wireless sensor networks. In: Proceedings of 2003 IEEE International Conference on Parallel Processing Workshops (2003)
177. Pietro, R.D., Mancini, L.V., Soriente, C., Spognardi, A., Tsudik, G.: Catch me (if you can): data survival in unattended sensor networks. In: Sixth Annual IEEE International Conference on Pervasive Computing and Communications (PerCom'08), pp. 185–194 (2008)
178. Piro, C., Shields, C., Levine, B.N.: Detecting the sybil attack in mobile ad hoc networks. In: Proceedings of IEEE 2nd International Conference on Security and Privacy in Communication Networks (SecureComm'06) (2006)
179. Poosala, V., Haas, P.J., Ioannidis, Y.E., Shekita, E.J.: Improved histograms for selectivity estimation of range predicates. SIGMOD Rec. **25**(2), 294–305 (1996)
180. Pottie, G.J., Kaiser, W.J.: Wireless integrated network sensors. Commun. ACM **43**(5), 51–58 (2000). http://doi.acm.org/10.1145/332833.332838
181. Priyantha, N.B., Chakraborty, A., Balakrishnan, H.: The cricket location-support system. In: Proceedings of the 6th Annual International Conference on Mobile Computing and Networking (MobiCom'00), pp. 32–43 (2000)
182. Rabaey, J.M., Ammer, M.J., da Silva, J.L., Patel, D., Roundy, S.: Picoradio supports ad hoc ultra-low power wireless networking. Computer **33**(7), 42–48 (2000). http://dx.doi.org/10.1109/2.869369
183. Rahman, S.M.M., Nasser, N., Inomata, A., Okamoto, T., Mambo, M., Okamoto, E.: Anonymous authentication and secure communication protocol for wireless mobile ad hoc networks. Secur. Commun. Netw. (Wiley) (2008)
184. Ranganathan, S., George, A.D., Todd, R.W., Chidester, M.C.: Gossip-style failure detection and distributed consensus for scalable heterogeneous clusters. Clust. Comput. **4**(3), 197–209 (2001). http://dx.doi.org/10.1023/A:1011494323443
185. Ren, K., Li, T., Wan, Z., Bao, F., Deng, R.H., Kim, K.: Highly reliable trust establishment scheme in ad hoc networks. Comput. Netw. **45**(6), 687–699 (2004). http://dx.doi.org/10.1016/j.comnet.2004.01.008
186. Rezvani, M., Ignjatovic, A., Bertino, E., Jha, S.: Secure data aggregation technique for wireless sensor networks in the presence of collusion attacks. IEEE Trans. Dependable Secur. Comput. **12**(1), 98–110 (2015)
187. Rivest, R., Adleman, L., Dertouzos, M.: On data banks and privacy homomorphisms. Found. Secur. Comput. pp. 169–179 (1978)
188. Rocha, B.P., Conti, M., Etalle, S., Crispo, B.: Hybrid static-runtime information flow and declassification enforcement. IEEE Trans. Inf. Forensics Secur. **8**(8), 1294–1305 (2013)
189. Roundy, S., Wright, P.K., Rabaey, J.: A study of low level vibrations as a power source for wireless sensor nodes. Comput. Commun. **26**(11), 1131–1144 (2003)

190. Roy, S., Conti, M., Setia, S., Jajodia, S.: Securely computing an approximate median in wireless sensor networks. In: Proceedings of the Forth International Conference on Security and Privacy in Communication Networks (SecureComm'08) (2008)
191. Roy, S., Conti, M., Setia, S., Jajodia, S.: Securely computing count and sum in wireless sensor networks. Technical Report, Center for Secure Information Systems—George Mason University (2008). http://mason.gmu.edu/sroy1/tr-count-sum.pdf
192. Roy, S., Conti, M., Setia, S., Jajodia, S.: Secure median computation in wireless sensor networks. Ad Hoc Netw. **7**(8), 1448–1462 (2009)
193. Roy, S., Conti, M., Setia, S., Jajodia, S.: Secure data aggregation in wireless sensor networks. IEEE Trans. Inf. Forensics Secur. **7**(3), 1040–1052 (2012)
194. Roy, S., Conti, M., Setia, S., Jajodia, S.: Secure data aggregation in wireless sensor networks: filtering out the attacker's impact. IEEE Trans. Inf. Forensics Secur. **9**(4), 681–694 (2014)
195. Roy, S., Setia, S., Jajodia, S.: Attack-resilient hierarchical data aggregation in sensor networks. In: Proceedings of the Fourth ACM Workshop on Security of Ad Hoc and Sensor Networks, pp. 71–82. ACM (2006)
196. Ruj, S., Nayak, A., Stojmenovic, I.: Pairwise and triple key distribution in wireless sensor networks with applications. IEEE Trans. Comput. **62**(11), 2224–2237 (2013)
197. Sanchez, D.S., Baldus, H.: A deterministic pairwise key pre-distribution scheme for mobile sensor networks. In: First IEEE/CreateNet International Conference on Security and Privacy for Emerging Areas in Communication Networks (SecureComm'05), pp. 277–288 (2005)
198. Sastry, N., Shankar, U., Wagner, D.: Secure verification of location claims. In: Proceedings of the 2nd ACM Workshop on Wireless Security (WiSe'03), pp. 1–10 (2003)
199. Sathish, R., Kumar, D.R.: Dynamic detection of clone attack in wireless sensor networks. In: Proceeding of the International Conference on Communication Systems and Network Technologies (CSNT'13), pp. 501–505. IEEE (2013)
200. Seshadri, A., Perrig, A., Doorn, L.V., Khosla, P.: SWATT: software-based attestation for embedded devices. In: Proceedings of the 2004 IEEE Symposium on Security and Privacy (S&P'04) (2004)
201. Shamir, A.: How to share a secret. Commun. ACM **22**(11), 612–613 (1979). http://doi.acm.org/10.1145/359168.359176
202. Shamir, A.: Identity-based cryptosystems and signature schemes. In: Proceedings of CRYPTO 84 on Advances in Cryptology, pp. 47–53 (1985)
203. Sharma, G., Mazumdar, R., Shroff, N.B.: Delay and capacity trade-offs in mobile ad hoc networks: a global perspective. In: 25th IEEE International Conference on Computer Communications (INFOCOM'06) (2006)
204. Shrivastava, N., Buragohain, C., Agrawal, D., Suri, S.: Medians and beyond: new aggregation techniques for sensor networks. In: Proceedings of the 2nd International Conference on Embedded Networked Sensor Systems (SenSys'04), pp. 239–249 (2004)
205. Smailagic, A., Siewiorek, D.P., Anhalt, J., Kogan, D., Wang, Y.: Location sensing and privacy in a context-aware computing environment. IEEE Wirel. Commun. **9**, 10–17 (2001)
206. Snekkenes, E.: Concepts for personal location privacy policies. In: Proceedings of the 3rd ACM Conference on Electronic Commerce (EC'01), pp. 48–57 (2001)
207. Song, H., Xie, L., Zhu, S., Cao, G.: Sensor node compromise detection: the location perspective. In: Proceedings of the 2007 International Conference on Wireless Communications and Mobile Computing (IWCMC'07), pp. 242–247 (2007)
208. Stankovic, J.A., Lu, C., Sha, L., Abdelzaher, T., Hou, J.: Real-time communication and coordination in embedded sensor networks. In: Proceedings of the IEEE, pp. 1002–1022 (2003)
209. Sterbenz, J.P.G., Krishnan, R., Hain, R.R., Jackson, A.W., Levin, D., Ramanathan, R., Zao, J.: Survivable mobile wireless networks: issues, challenges, and research directions. In: Proceedings of the 1st ACM Workshop on Wireless Security (WiSe'02), pp. 31–40 (2002)
210. Striki, M., Baras, J., Manousakis, K.: A robust, distributed TGDH-based scheme for secure group communications in MANET. In: Proceedings of the 41th IEEE International Conference on Communications (ICC'06), pp. 2249–2255 (2006)

211. Sun, K., Ning, P., Wang, C.: Fault-tolerant cluster-wise clock synchronization for wireless sensor networks. IEEE Trans. Dependable Secur. Comput. **2**(3), 177–189 (2005)
212. Tanachaiwiwat, S., Dave, P., Bhindwale, R., Helmy, A.: Poster abstract secure locations: routing on trust and isolating compromised sensors in location-aware sensor networks. In: Proceedings of the 1st International Conference on Embedded Networked Sensor Systems (SenSys'03), pp. 324–325 (2003)
213. Tanachaiwiwat, S., Dave, P., Bhindwale, R., Helmy, A.: Location-centric isolation of misbehavior and trust routing in energy-constrained sensor networks. In: IEEE Workshop on Energy-Efficient Wireless Communications and Networks (EWCN), in Conjunction with IEEE (IPCCC'04), pp. 14–17 (2004)
214. Vasudevan, S., DeCleene, B., Immerman, N., Kurose, J., Towsley, D.: Leader election algorithms for wireless ad hoc networks. DARPA Inf. Surviv. Conf. Expos. (DISCEX'03), p. 261 (2003)
215. Čapkun, S., Buttyán, L., Hubaux, J.P.: Sector: secure tracking of node encounters in multi-hop wireless networks. In: Proceedings of the 1st ACM Workshop on Security of Ad Hoc and Sensor Networks (SASN'03), pp. 21–32. ACM, New York (2003)
216. Čapkun, S., Hubaux, J.: Secure positioning in wireless networks. In: Proceedings of the IEEE Journal on Selected Areas in Communication: Special Issue on Security Wireless Ad Hoc Network **24**(2), 221–232 (2006)
217. Čapkun, S., Hubaux, J.P.: Secure positioning of wireless devices with application to sensor networks. In: Proceedings of the 24th Annual Joint Conference of the IEEE Computer and Communications Societies (INFOCOM'05), pp. 1917–1928 (2005)
218. Čapkun, S., Hubaux, J.P., Buttyán, L.: Mobility helps security in ad hoc networks. In: Proceedings of the 4th ACM International Symposium on Mobile Ad Hoc Networking and Computing (MobiHoc'03), pp. 46–56 (2003)
219. Čapkun, S., Čagalj, M.: Integrity regions: authentication through presence in wireless networks. In: Proceedings of the 5th ACM Workshop on Wireless Security (WiSe'06), pp. 1–10 (2006)
220. Wagner, D.: Resilient aggregation in sensor networks. In: Proceedings of the 2nd ACM Workshop on Security of Ad Hoc and Sensor Networks (SASN'04), pp. 78–87 (2004)
221. Walters, J.P., Liang, Z., Shi, W., Chaudhary, V.: Wireless sensor network security: a survey. In: Xiao, Y. (ed.) Security in Distributed, Grid, Mobile, and Pervasive Computing, pp. 367–417. Auerbach Publications, CRC Press, Boca Raton (2007)
222. Wander, A., Gura, N., Eberle, H., Gupta, V., Shantz, S.C.: Energy analysis of public-key cryptography for wireless sensor networks. In: Proceedings of the Third Annual IEEE International Conference on Pervasive Computing and Communications Workshops (PERCOMW'05) (2005)
223. Wang, W., Bhargava, B.: Visualization of wormholes in sensor networks. In: Proceedings of the 3rd ACM Workshop on Wireless Security (WiSe'04), pp. 51–60 (2004)
224. Wang, X., Gu, W., Chellappan, S., Schosek, K., Xuan, D.: Lifetime optimization of sensor networks under physical attacks. In: Proceedings of IEEE International Conference on Communications (ICC'05) (2005)
225. Wang, X., Gu, W., Schosek, K., Chellappan, S., Xuan, D.: Sensor network configuration under physical attacks. In: Networking and Mobile Computing, Third International Conference (ICCNMC'05), pp. 23–32 (2005)
226. Warneke, B., Last, M., Liebowitz, B., Pister, K.S.J.: Smart dust: communicating with a cubic-millimeter computer. Computer **34**(1), 44–51 (2001). http://dx.doi.org/10.1109/2.963443
227. Watro, R., Kong, D., Cuti, S., Gardiner, C., Lynn, C., Kruus, P.: Tinypk: securing sensor networks with public key technology. In: Proceedings of the 2nd ACM Workshop on Security of Ad Hoc and Sensor networks (SASN'04), pp. 59–64 (2004)
228. Weiser, M.: Some computer science issues in ubiquitous computing. Commun. ACM **36**(7), 75–84 (1993). http://doi.acm.org/10.1145/159544.159617
229. Williams, B., Camp, T.: Comparison of broadcasting techniques for mobile ad hoc networks. In: Proceedings of the 3rd ACM International Symposium on Mobile Ad Hoc Networking and Computing (MobiHoc'02), pp. 194–205 (2002)

230. Wood, A.D., Stankovic, J.A.: Denial of service in sensor networks. Computer **35**(10), 54–62 (2002). http://dx.doi.org/10.1109/MC.2002.1039518

231. Xi, Y., Schwiebert, L., Shi., W.: Preserving privacy in monitoring-based wireless sensor networks. In: Proceedings of the 2nd International Workshop on Security in Systems and Networks (SSN'06) (2006)

232. Xie, W., Wang, L., Wang, M.: A bloom filter and matrix-based protocol for detecting node replication attack. J. Netw. **9**(6), 1471–1476 (2014)

233. Yan, Z., Zhang, P., Virtanen., T.: Trust evaluation based security solutions in ad-hoc networks. In: Proceedings of the Seventh Nordic Workshop on Security IT Systems (NordSec'03) (2003)

234. Yang, Y., Shao, M., Zhu, S., Urgaonkar, B., Cao, G.: Towards event source unobservability with minimum network traffic in sensor networks. In: Proceedings of the First ACM Conference on Wireless Network Security (WiSec'08), pp. 77–88 (2008)

235. Yang, Y., Wang, X., Zhu, S., Cao, G.: SDAP: a secure hop-by-hop data aggregation protocol for sensor networks. In: Proceedings of the 7th ACM International Symposium on Mobile Ad Hoc Networking and Computing (MobiHoc'06), pp. 356–367 (2006)

236. Yao, A.: Protocols for secure computations. In: Proceedings of the Symposium on Foundations of Computer Science (FOCS'82), pp. 160–164 (1982)

237. Yoon, J., Liu, M., Noble, B.: Random waypoint considered harmful. In: Proceeding of the 22th Annual Joint Conference of the IEEE Computer and Communications Societies (INFO-COM'03), pp. 1312–1321 (2003)

238. Zanin, G., Pietro, R.D., Mancini, L.V.: Robust RSA distributed signatures for large-scale long-lived ad hoc networks. J. Comput. Secur. **15**(1), 171–196 (2007)

239. Zeng, W., Lin, Y., Yu, J., He, S., Wang, L.: Privacy-preserving data aggregation scheme based on the p-function set in wireless sensor networks. Ad-hoc Sens. Wirel. Netw. **21**(1–2), 21–58 (2014)

240. Zhang, L., Zhang, H., Conti, M., Di Pietro, R., Jajodia, S., Mancini, L.V.: Preserving privacy against external and internal threats in wsn data aggregation. Telecommun. Syst. **52**(4), 2163–2176 (2013)

241. Zhang, Q., Yu, T., Ning, P.: A framework for identifying compromised nodes in wireless sensor networks. ACM Trans. Inf. Syst. Secur. **11**(3), 1–37 (2008)

242. Zhang, X., Chen, H., Wang, K., Peng, H., Fan, Y., Li, D.: Rotation-based privacy-preserving data aggregation in wireless sensor networks. In: Proceeding of the IEEE International Conference on Communications (ICC'14), pp. 4184–4189. IEEE (2014)

243. Zhauniarovich, Y., Russello, G., Conti, M., Crispo, B., Fernandes, E.: Moses: supporting and enforcing security profiles on smartphones. IEEE Trans. Dependable Secur. Comput. **11**(3), 211–223 (2014)

244. Zhu, B., Addada, V.G.K., Setia, S., Jajodia, S., Roy, S.: Efficient distributed detection of node replication attacks in sensor networks. In: 23rd Annual Computer Security Applications Conference (ACSAC'2007), pp. 257–266 (2007)

245. Zhu, B., Setia, S., Jajodia, S., Roy, S., Wang, L.: Localized multicast: efficient and distributed replica detection in large-scale sensor networks. IEEE Trans. Mob. Comput. **9**(7), 913–926 (2010)

246. Zhu, H., Bao, F., Deng, R.H., , Kim., K.: Computing of trust in wireless networks. In: Proceedings of 60th IEEE Vehicular Technology Conference (VTC'05) (2005)

247. Zhu, S., Setia, S., Jajodia, S.: LEAP: efficient security mechanisms for large-scale distributed sensor networks. In: Proceedings of the 10th ACM Conference on Computer and Communications Security (CCS'03), pp. 62–72 (2003)

248. Zhu, S., Setia, S., Jajodia, S.: LEAP: efficient security mechanisms for large-scale distributed sensor networks. In: Proceedings of the 10th ACM Conference on Computer and Communications Security (CCS'03), pp. 62–72 (2003)

249. Zhu, S., Xu, S., Setia, S., Jajodia, S.: Establishing pair-wise keys for secure communication in ad hoc networks: a probabilistic approach. In: Proceedings of the 11th IEEE International Conference on Network Protocols (ICNP'03), p. 326 (2003)

Printed in the United States
By Bookmasters